SpringerBriefs in Education

More information about this series at http://www.springer.com/series/8914

Azra Moeed

Science Investigation

Student Views about Learning, Motivation and Assessment

 Springer

Azra Moeed
Faculty of Education
Victoria University of Wellington
Karori, Wellington
New Zealand

ISSN 2211-1921   ISSN 2211-193X   (electronic)
SpringerBriefs in Education
ISBN 978-981-287-383-5   ISBN 978-981-287-384-2   (eBook)
DOI 10.1007/978-981-287-384-2

Library of Congress Control Number: 2014960204

Springer Singapore Heidelberg New York Dordrecht London

Printed on acid-free paper

Springer Science+Business Media Singapore Pte Ltd. is part of Springer Science+Business Media
(www.springer.com)

Acknowledgments

I would like to thank Dr. Janet Davies for inspiring me to do this research and for her guidance. My thanks to Dr. Linda Leach and Dr. Stephanie Doyle for their supervision and to Dr. Abdul Moeed for his support and for sustaining me while doing it. Thanks to Susan Kaiser for editorial support and for feedback on earlier drafts.

Contents

Chapter 1
Introduction

Learning science involves developing conceptual and procedural understanding as well as understanding of the Nature of Science. Internationally, the focus of science education at the start of this millennium has been for all students to be scientifically literate. It is argued that such a citizen will be able to make informed decisions about scientific developments and rapidly growing new technologies that are likely to impact on their everyday lives.

Globally, schools and teachers are grappling with this need of science *education for all* and preparing students who will continue to study science at senior secondary school and then be inspired to study science with a view to becoming creative and innovative scientists contributing to the progress of their respective countries. Added to this are the short term goals for schools to prepare students to be successful in their school science education through achieving in their examinations.

Scientists investigate questions that are of interest to them. They have subject specific knowledge and conceptual understandings. By working in their field they also have considerable skills and procedural knowledge to know how science works. However, when students ask a question they seldom demonstrate conceptual or procedural knowledge and may not have an in depth understanding of how science works to plan and carry out an investigation.

The challenge, then, for school science is to provide learning experiences that allow students to explore, learn science ideas, learn the many approaches to science investigation, and develop an understanding of the nature of science. It is when students are experiencing science and beginning to understand science ideas and how to do science that they will be able to ask questions and have some idea as to how they will find the answer to their questions. Schools need to start this early and carry it through the compulsory years of education.

Research into students' views about their learning through practical work in general, and science investigation in particular, have often relied on surveys (Bennett et al. 2005; Holstermann et al. 2010; Toplis 2012). Student engagement in practical work is an under researched area (Alsop 2005). More recently,

© The Author(s) 2015
A. Moeed, *Science Investigation*, SpringerBriefs in Education,
DOI 10.1007/978-981-287-384-2_1

Abrahams and Millar (2008) have made a substantial contribution through researching the effectiveness of practical work, and Abrahams (2009) explored the affective value of practical work in schools. The potential for raising secondary school students' science achievement and its long-term effects through engagement in Cognitive Acceleration through Science Education (CASE) has been researched at King's College in the UK (Adey and Shayer 2002; Adey et al. 1995). The *Thinking Science Australia* programme has reported significant gains in achievement of students from low socio-economic areas (Oliver et al. 2012).

Hume and Coll (2008) have investigated student experience of science investigation in a case study, and report that students were developing a very narrow view of science investigation as fair testing. Their later paper (Hume and Coll 2009) focuses on assessment of science investigation. A limited number of studies have explored students' thoughts about learning, motivation to learn, and internal assessment of science investigation (Lyons 2006; Moeed 2010). The present study is about connectedness and reports on the complexity of the laboratory environment, student motivation to learn, learning and assessment of science investigation and its affordances and constraints as experienced and articulated by students.

Internationally, learning to investigate is a mandatory requirement of most secondary school curricula. We have moved on from science being *hands-on* to expect it to be *minds-on*. If we want students to *do* science, we want them to *learn* science. This book illuminates what a class of 15-year-old students in a typical coeducational, middle size school in a middle socio-economic community say about their experiences of investigating in their science classes. The book presents the research findings of an interpretive case study that explored learning science investigation (scientific inquiry) from students' perspectives and what these students found motivational, what they learnt, and how they were assessed. It attempts to explain the connectedness between motivation to learn, learning, and assessment of science investigation.

1.1 Nature of Science Investigation

The term 'science investigation' (or 'investigative work in science') in this book refers to an activity in which students try to answer a question or solve a problem. In doing this, students have some freedom to choose the methods and the equipment they will use, the data they will collect, and the ways in which they will interpret the data. The question or problem itself may be identified by the students, or given by the teacher, or arrived at through teacher-student discussion.

Millar (2012) describes an 'investigation' as:

> an activity requiring identification of a question, using both conceptual and procedural knowledge in planning and carrying out the investigation, gathering, processing, and interpreting data and drawing conclusions based on evidence. Ideally, the process is iterative and the student has some choice in what they want to investigate. (p. 2)

Science investigation is a commonly used term in the UK, Australia and New Zealand, while the term *science inquiry* is used in the USA. Investigations are mostly, but not always, practical in nature (Gott and Duggan 1996). A practical science investigation is one form of practical work, and is different to 'recipe following' practical activities in which students simply have to follow a set of written or oral instructions.

Students gain most from science investigation when they have the opportunity to predict what is likely to happen and explain reasons, make observations, draw evidence-based conclusions and discuss their theories, before, during, and after conducting the investigation (Patrick and Yoon 2004). Learning investigation needs to be seen as a recursive process rather than a constrained procedure.

Focussing on science investigation to develop conceptual understanding, science educators propose that carrying out a complete investigation of this kind enables students not just to *do science* but also to learn the science concepts and understand the nature of science (Hodson 1990; Roberts and Gott 2006).

In response to a loss of interest in taking science in secondary schools in Australia, Tytler (2007) suggested a "re-imagined curriculum" with thoughts of what investigative science should look like:

> Science investigations should be more varied, with explicit attention paid to investigative principles. Investigative design should encompass a wide range of methods and principles of evidence including sampling, modelling, field-based methods, and the use of evidence in socio-scientific issues. Investigations should frequently flow from students' own questions. Investigations should exemplify the way ideas and evidence interact in science. (p. 64)

In science classrooms, practical work, experiments, inquiry and investigation are often used interchangeably. The following section defines the different terms in order to clarify the confusion.

1.2 Sorting Out the Terminology

The preference for science investigation rather than scientific inquiry is to steer away from misconceptions about inquiry learning, and teaching as inquiry that have different meanings in both literature and practice:

> 'Teaching as inquiry' is when teachers inquire into: what is most important; what strategies or approaches are most likely to work; and the impact of teaching on students. Whereas 'inquiry learning' is just one approach teachers might use (but don't have to) in which students learn about learning, investigation and research as they explore topics of interest. (Ministry of Education 2014, p. 35)

Scientific inquiry is the way scientists carry out their research which leads to scientific knowledge being generated and accepted. It is how scientists develop scientific knowledge through creatively applying generic science processes, science content, and critical thinking (Lederman 2009). The difference between

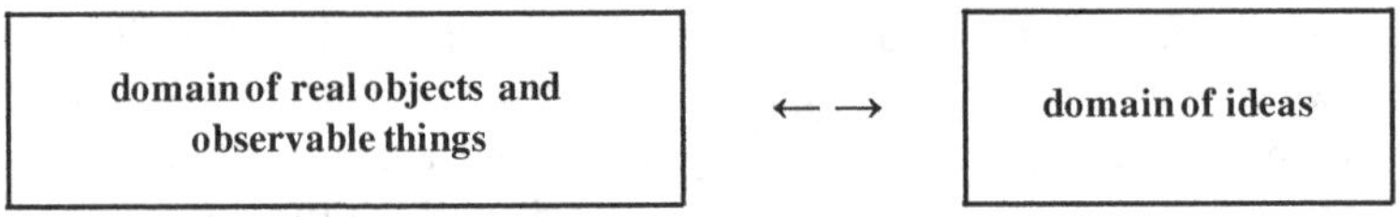

| domain of real objects and observable things | ← → | domain of ideas |

Fig. 1.1 Practical work: linking two domains of knowledge (Millar 2004, p. 8)

doing inquiry and *understanding about the nature of scientific inquiry* is beyond the scope of this book and it has been most eloquently presented by Lederman et al. (2014).

1.2.1 Practical Work

Practical work is any science teaching and learning activity where students observe or manipulate objects individually or in small groups. Practical work has generally been called hands-on activities or doing science; it may or may not take place in the laboratory (Millar 2004). Most practical work undertaken in secondary school classrooms falls under this category. If students are engaged in practical work then the focus ought to be on learning, and it is noteworthy that much of the learning from practical activities takes place in the discussion that follows. For Millar, the role of practical work in the teaching and learning of science content is to help students make links between two "domains" of knowledge as shown in Fig. 1.1.

1.2.2 Experiments

School science experiments involve replication of a piece of practical work already carried out by scientists, either recently or long ago. Some argue that the term "experiment" should be "reserved.... for true Popper-type experiments where an abstract high level theory, model of law is put to a cleverly devised check" (Finch and Coyle 1988, p. 45). Abrahams and Millar (2008) describe an experiment as "a planned intervention to test a prediction derived from a theory, hypothesis" (p. 1947). Thus, an experiment in school science is an intervention or manipulation of objects that helps in understanding the material world. It may involve replication of an activity to confirm a theory learnt and therefore be a standalone activity; for example, if a metal rod is heated, it expands. Through doing this experiment, a student can confirm the theory that metals expand when heated. It can also be said that the intervention of heating the metal can prove the prediction that metals expand when heated.

   Alternatively, an experiment can be carried out within or as part of an investigation. If the student is to carry out an investigation, it would involve trying out their plan to see if it *works* and to evaluate and make decisions about proceeding with the investigation. This *experiment* could lead to altering the method and, in this case, the experiment would be within an investigation.

1.2.3 Scientific Method

Scientific method is described as a series of steps that include: observing, defining the problem, gathering reliable data, selecting an appropriate hypothesis to explain the data, planning, carrying out experiments or observations to test the hypothesis, and drawing a conclusion in support or otherwise of the tested hypothesis (Moeed 2013; Wong et al. 2008). Most critics of a scientific method consider following ordered steps that are inadequate as a description of scientific practice or as a guide to instruction (Hodson 1996; Lederman 1998; Tang et al. 2010).

Scientists carry out investigations in a variety of ways; they may not follow a particular method but many different methods depending on the nature of the investigation. Most scientists agree that there is no single 'scientific method' and that a simplistic representation of how scientists investigate is misguided. Davies (2005) argues that there are differences "in practice between science disciplines and between individual topics within disciplines" (p. 3). She attributes these differences to the nature of the questions asked, and the theories, methods and equipment used. The scientific method may be useful as a framework for reporting the results of an investigation but the neatness of the research report does not reflect the messiness of the process (Mahootian and Eastman 2008). The notion of a framework to report findings appears to be aligned with the manner in which the scientific method is set out in the introductory chapters of science textbooks used by students in junior science classes (Wong et al. 2008). An additional aspect of this commonly used scientific method is that making a hypothesis is considered by many students and teachers to be an essential part of this linear method of scientific knowledge generation (Wong et al. 2008).

1.2.4 Approaches to Investigation

Watson et al. (1999) proposed six different types of investigations for school science that cover a range of skills and give students the opportunity to understand science ideas and how science works. Along with the types of investigations, they succinctly explained what students will be doing when carrying them out, and provided examples:

1. Classifying and identifying: this involves grouping things and looking for patterns either through experimenting or from information from databases. The students could look at a collection of shellfish and group them as having one or two shells.

2. Fair testing: these apply to situations where the students observe the relationships between two variables. One variable is changed (independent variable) and the other factors are controlled. An examples here could be investigating the need for both water and oxygen for iron to rust.

3. Pattern seeking: considering large amounts of data where variables cannot be easily controlled. They recognise the importance of having a large sample to ensure that any conclusions that are drawn are significant. They are also looking for cause and effect. Examples would include biological surveys where there will be variations within a population.

4. Investigating models: testing to see if the models explain certain phenomena. Students have to look at the explanations presented by one scientist to another and on the basis of the evidence presented put forth their own ideas.

5. Exploring: students make a study of a change over a period of time. This could include observations of life present in rock pools in high, mid, and low tide areas. Students raise questions from their observations.

6. Making things or developing systems: students apply the knowledge and procedures they have learnt in science to develop artefacts for specific use.

These six types of investigations are promoted in New Zealand through the use of textbooks and workbooks that set out tasks students can carry out (Abbott et al. 2005). Some of the examples listed above are also used in the list of learning experiences in *Science in the New Zealand Curriculum* (Ministry of Education 1993). The same six types of investigation are promoted in the *New Zealand Curriculum* (Ministry of Education 2007) as approaches to investigation, and textbooks and workbooks used in New Zealand set out tasks students can carry out (Abbott et al. 2005).

Fair testing was important in research by Moeed (2010) as it was emphasised as a type of investigation in *Science in the New Zealand Curriculum* (Ministry of Education 1993) and it was the type of investigation that was assessed through Achievement Standard AS1.1, New Zealand Qualifications Authority (NZQA 2005). Fair testing types of investigation have been criticised as linear and sequential (Hume and Coll 2008; Moeed 2010). Watson et al. (1999) found that in the UK the national curricula have an "over-heavy" emphasis on fair testing and that this is detrimental to other kinds of investigation.

1.3 The Context of the Study

This brief presents the findings in relation to students' views that were an aspect of a research project that investigated teaching, learning, motivation to learn, and assessment of science investigation in year 11 in New Zealand. Within a qualitative research paradigm, the research took a case study approach and studied the phenomenon of science investigation. Qualitative research is interpretive, has a naturalistic orientation and allows the use of multiple methods (Denzin and Lincoln 2005). The intention was to understand investigation, the phenomenon of interest through those who experience it in their unique contexts and through the interactions that take place in that setting (Merriam 1998).

An in depth, nested case study of a science class and their teacher in a large, urban, coeducational secondary school is presented. The data were gathered through Science Laboratory Environment Inventory, classroom observations, focus group interviews, informal conversations with the students during the lesson, observation of the assessment process, and document analysis. For detailed methodology please read Moeed (2010).

1.4 The Research Questions

The research focussed on the following questions:

1. How does carrying out science investigation relate to students' learning and motivation to learn?
2. What affect does internal assessment have on year 11 students' learning and motivation to learn?
3. How does the type of investigation relate to students' learning and motivation to learn?

1.5 Outline of This Book

This Brief tells the story of a class of year 11 students—what they learnt, what motivated them to learn, how they prepared for assessment—and comprises five chapters. This chapter provides a general introduction to science investigation, and explores the literature to define relevant terminology and types of investigations. Chapter 2 sets the scene and the laboratory environment where the participants of this research learnt science investigation. Research findings in relation to what the laboratory environment was like and what the students would have preferred it to be like are presented. Illustrative examples of student learning through science investigation are also stated. Chapter 3 gives a brief review of selected relevant literature about motivation to learn and findings about student engagement, and their views about motivation to learn science investigation are discussed. Chapter 4 is devoted to assessment of science investigation. The chapter starts with assessment literature and describes student experience of formative and summative assessment. Chapter 5 presents the integrated results and compares and contrasts data from the multiple sources as well as discussing the emerging themes. The focus is on the complexity of learning to investigate and on the three aspects of learning, motivation, and assessment. It also includes the conclusions in relation to the research questions. The Brief ends with final thoughts for teaching science investigation.

References

Abbott, G., Cooper, G., & Hume, A. (2005). Year 11 science (NCEA level 1) workbook. Hamilton, New Zealand: ABA.

Abrahams, I. (2009). Does practical work really motivate? A study of the affective value of practical work in secondary school science. *International Journal of Science Education, 31*(17), 2335–2353.

Abrahams, I., & Millar, R. (2008). Does practical work really work? A study of the effectiveness of practical work as a teaching and learning method in school science. *International Journal of Science Education, 30*(14), 1945–1969. doi:10.1080/09500690701749305.

Adey, P., & Shayer, M. (2002). Cognitive acceleration comes of age. In M. Shayer & P. Adey (Eds.), *Learning intelligence* (pp. 1–17). Buckingham: Open University Press.

Adey, P., Shayer, M., & Yates, C. (1995). *Thinking science: Student and teachers' materials for the CASE intervention* (2nd ed.). London: Macmillan.

Alsop, S. (2005). Bridging the Cartesian divide: Science education and affect. *Beyond Cartesian dualism* (pp. 3–16). Netherlands: Springer.

Bennett, J., Lubben, F., Hogarth, S., & Campbell, B. (2005). Systematic reviews of research in science education: Rigour or rigidity? *International Journal of Science Education, 27*(4), 387–406.

Davies, J. (2005, July). Learning about science: Towards school science as 'a way of being in the world'. In *Paper presented to international history and philosophy of science teaching annual conference*. UK: University of Leeds.

Denzin, N. K., & Lincoln, Y. S. (Eds.). (2005). *The Sage handbook of qualitative research* (3rd ed.). Thousand Oaks, CA: Sage.

Finch, I., & Coyle, M. (1988). Teacher education project, INSET secondary science: Teachers' notes. In R. Millar (Ed.), (1989), *Doing science: Images of science in science education*. London: The Falmer Press.

Gott, R., & Duggan, S. (1996). Practical work: Its role in the understanding of evidence in science. *International Journal of Science Education, 18*(7), 791–806.

Hodson, D. (1990). A critical look at practical work in school science. *School Science Review, 71*(256), 33–40.

Hodson, D. (1996). Laboratory work as scientific method: Three decades of confusion and distortion. *Journal of Curriculum Studies, 28*(2), 115–135.

Holstermann, N., Grube, D., & Bögeholz, S. (2010). Hands-on activities and their influence on students' interest. *Research in Science Education, 40*(5), 743–757.

Hume, A., & Coll, R. (2008). Student experiences of carrying out a practical science investigation under direction. *International Journal of Science Education, 30*(9), 1201–1228.

Hume, A., & Coll, R. (2009). Assessment of learning, for learning, and as learning: New Zealand case studies. *Assessment in Education Principles Policy and Practice, 16*(3), 269–290.

Lederman, N. (1998). The state of science education: Subject matter without context. *Electronic Journal of Science Education, 3*, 1–11.

Lederman, J. S. (2009). Development of a valid and reliable protocol for the assessment of early childhood students' conceptions of nature of science and scientific inquiry (Doctoral dissertation, Curtin University of Technology).

Lederman, J. S., Lederman, N. G., Bartos, S. A., Bartels, S. L., Meyer, A. A., & Schwartz, R. S. (2014). Meaningful assessment of learners' understandings about scientific inquiry: The views about scientific inquiry (VASI) questionnaire. *Journal of Research in Science Teaching, 51*(1), 65–83.

Lyons, T. (2006). Different countries, same science classes: Students' experiences of school science in their own words. *International Journal of Science Education, 28*(6), 591–613.

Mahootian, F., & Eastman, T. E. (2008). Complementary frameworks of scientific inquiry: Hypothetico-deductive, hypothetico-inductive, and observational-inductive. *World Futures, 65*(1), 61–75. doi:10.1080/02604020701845624.

Merriam, S. B. (1998). *Qualitative research and case study applications in education.* San Francisco: Jossey-Bass.

Millar, R. (2004). The role of practical work in the teaching and learning of science. In *Paper presented for the meeting of high school science laboratories: Role and vision.* Washington, DC: National Academy of Sciences.

Millar, R. (Ed.). (2012). *Doing science (RLE Edu O): Images of science in science education.* New York: Routledge.

Ministry of Education (1993). *Science in the New Zealand Curriculum.* Wellington: Learning Media.

Ministry of Education (2014). Teaching as inquiry: Ask yourself the big question. *New Zealand Education Gazette, 96*(6), Retrieved April 21, 2014, from http://www.edgazette.govt.nz/Articles/Article.aspx?ArticleId=7880.

Ministry of Education (2007). *The New Zealand Curriculum.* Wellington: Learning Media.

Moeed, H. A. (2010). Science investigation in New Zealand secondary schools: Exploring the links between learning, motivation and internal assessment in Year 11. Unpublished doctoral dissertation, Victoria University of Wellington, New Zealand.

Moeed, A. (2013). Science investigation that best supports student learning: Teachers' understanding of science investigation. *International Journal of Environmental and Science Education, 8*(4), 537–559.

Oliver, M., Venville, G., & Adey, P. (2012). Effects of a cognitive acceleration programme in a low socioeconomic high school in regional Australia. *International Journal of Science Education, 34*(9), 1393–1410.

Patrick, H., & Yoon, C. (2004). Early adolescents' motivation during science investigation. *The Journal of Educational Research, 97*(6), 319–328.

Roberts, R., & Gott, R. (2006). Assessment of performance in practical science and pupil attributes. *Assessment in education, 13*(01), 45–67.

Tang, X., Coffey, J. E., Elby, A., & Levin, D. M. (2010). The scientific method and scientific inquiry: Tensions in teaching and learning. *Science Education, 94*(1), 29–47. doi:10.1002/sce.20366.

Toplis, R. (2012). Students' views about secondary school science lessons: The role of practical work. *Research in Science Education, 42*(3), 531–549.

Tytler, R. (2007). Re-imagining science education: Engaging students in science for Australia's future.

Watson, R., Goldsworthy, A., & Wood-Robinson, V. (1999). What is not fair with investigations? *School Science Review, 80*(292), 101–106.

Wong, S. L., Hodson, D., Kwan, J., & Yung, B. H. W. (2008). Turning crisis into opportunity: Enhancing student teachers' understanding of the nature of science and scientific inquiry through a case study of the scientific research in severe acute respiratory syndrome. *Science and Education, 30*(11), 1417–1439.

Chapter 2
Science Laboratory Learning Environment, and Learning

2.1 Science Laboratory and Learning

Internationally, practical work in a laboratory setting is considered to be a preferred learning environment for school science. Millar (2010) argues that if science education aims for students to understand the natural world and how it functions, then learners need to experience and observe the relevant science phenomena. Millar argues that through observation and manipulation of objects students are more likely to make the links between the domains of objects and ideas. Other science education researchers have argued that many benefits accrue from engaging students in laboratory activities in science (Dkeidek et al. 2012; Hofstein 2004; Hofstein et al. 2008; Woolnough 1991). Dkeidek et al. (2012) consider science laboratories provide an ideal environment for students to work cooperatively and investigate scientific phenomena. Assertions are made about the benefits of laboratory work, including fostering positive attitudes and interest in science (Hoffstein and Lunetta 1982; Luketic and Dolan 2013). Hodson (1990) suggests that the main reasons given by teachers in support of practical work include: motivational benefits (interest and enjoyment); development of skills and science knowledge; learning about and following scientific method; and developing scientific attitudes. However, there is considerable doubt about the effectiveness of laboratory work in school science (Abrahams and Millar 2008). In the literature, practical work and laboratory work are used interchangeably, although Millar (2010) argues that practical work can also take place outside the laboratory. However, in most high schools it usually occurs in the laboratory.

Osborne (1998) argues that laboratory work has a limited role and little educational value in science learning. Hodson (1991) asserts: "despite its often massive share of curriculum time, laboratory work often provides little of real educational value. As practiced in many countries, it is ill-conceived, confused and unproductive". Hodson goes on to say, "For many children, what goes on in the laboratory

© The Author(s) 2015
A. Moeed, *Science Investigation*, SpringerBriefs in Education,
DOI 10.1007/978-981-287-384-2_2

contributes little to their learning of science or to their learning about science and its methods" (p. 176). Abrahams (2009) has challenged the claim that laboratory work is motivational. In recent times, the attention has turned to finding out what students say about their learning.

Toplis (2012) investigated students' views about teaching and learning through practical work in the science laboratory. When given the voice in Toplis's study, students said laboratory work was important for interest and activity, and it gave them autonomy and an opportunity to participate. Students did not like transmission of knowledge from the teacher, rote learning and recall of facts. Rudduck and McIntyre (2007), in their research, found that students described good lessons for learning when: there was a lack of tedium; learning was meaningful; they were able to work together; and they had autonomy. It appears that when students have a choice, the learning environment that gives them autonomy and freedom to draw upon the ideas of their peers along with engaging in interesting tasks, they learn science.

2.2 Science Laboratory Environment Inventory

The SLEI was developed and validated by Fraser et al. (1995). This instrument is especially suited to assessing the environment of science classes because of the importance of laboratory settings in science education. The inventory comprises 35 items, each of which is judged on a scale of 1–5. This SLEI has an actual and a preferred version of the learning environment. In the actual version the students respond to the questions by selecting options that indicate how things happen in their laboratory class. In the preferred version they choose responses that indicate what they wished the classroom environment to be. The inventory has five scales: student cohesiveness, open-endedness; integration; rule clarity; and material environment. The SLEI was field tested and validated with 5,447 students from 269 classes in seven countries, including Australia. Fraser et al. (1995) used individual students as a unit of analysis and reported the internal consistency (alpha reliability) and discriminant validity (mean correlation of a scale with the other four scales). The statistics are reported in Table 2.1.

In the alpha reliability data, the numbers in the left-hand columns, are all relatively high; they show how the items in each set are internally correlated (e.g., all those items that measure actual student cohesiveness are quite highly (0.80) interrelated). The numbers in the right hand-columns are all relatively low; they show that the sets themselves are not highly correlated, for example, actual student cohesiveness has a low average correlation (0.31) with all the other scales. The description of each scale and an illustrative sample item are presented in Table 2.2 which is from Hofstein (2004, p. 355).

The SLEI has been used globally. Fraser and McRobbie (1995) investigated students' perceptions of their science learning environment in six countries including; UK, Nigeria, Australia, Israel, USA, and Canada. Hofstein et al. (2001) used the SLEI to compare the learning environments for students who engaged in an inquiry approach with a control group. They reported that the inquiry group found

Table 2.1 Internal consistency (Cronbach alpha reliability) and discriminant validity (mean correlation with other scales) for actual and preferred versions for a cross-validation sample for class mean as a unit of analysis

Scale	Alpha reliability		Mean correlation with other scales	
	Actual	Preferred	Actual	Preferred
Student cohesiveness	0.80	0.82	0.31	0.31
Open-endedness	0.80	0.70	0.25	0.15
Integration	0.91	0.92	0.44	0.36
Rule clarity	0.76	0.80	0.43	0.35
Material environment	0.74	0.85	0.34	0.40

Note Table from Fraser et al. (1995, p. 15)

Table 2.2 Descriptive information and sample items for each scale of SLEI

Scale name	Description	Sample item
Student cohesiveness	Extent to which students know, help and are supportive of one another	Members of this laboratory class help one another
Open-endedness	Extent to which the laboratory activities emphasise an open-ended, divergent approach to experimentation	In our laboratory sessions, different students do different experiments
Integration	Extent to which the laboratory activities are integrated with non-laboratory and theory classes	We use the theory from our regular class sessions during laboratory activities
Rule clarity	Extent to which behaviour in the laboratory is guided by formal rules	There is a recognised way of doing things safely in this laboratory
Material environment	Extent to which the laboratory equipment and materials are adequate	The laboratory has enough room for individual or group work

the actual learning environment more aligned with their preferred environment than the control group. More recently, Dkeidek et al. (2012) used a SLEI (translated in Arabic and validated) to compare the effect of culture on Jewish and Arab students' perceptions of their learning environments. Dkeidek et al. (2012) found that the responses were different in the pre-inquiry phase and similar in the categories that were measured during the inquiry phase.

2.3 Methodology

Although this SLEI was not tested in New Zealand, it was considered appropriate because the instrument was using items that were relevant to science learning in New Zealand laboratories. Both actual and preferred versions of the SLEI were administered. Students took approximately 25 min to complete the actual version and about 20 min for the preferred version. The students selected from five scoring options ranging from 1 (low) to 5 (high). The negative items were reversed before adding the responses for the five scales (see Table 2.1). A paired sample t test was applied for both actual and preferred items on each of the scales (see Table 2.3). The five scales used were open-endedness (items 2, 7, 12, 17, 22, 27, 32), material environment (items 5, 10, 15, 20, 25, 30, 35), integration (items 3, 8, 13, 18, 23, 28, 33), rule clarity (items 4, 9, 14, 19, 24, 29, 34), and cohesiveness (items 1, 6, 11, 16, 21, 26, 31). Open-endedness items relate to student control over the design and implementation of practical work, and the choice they had or would like to have had in investigating what they wanted to investigate. Material environment items were related to the physical aspects such as whether they had the equipment available to do the practical work and if it was in good repair. Integration items were about the relationship between the concepts they had learnt and the practical work they were doing. Rule clarity was about the parameters within which they had to work, for example how they understood the safety requirement. A cohesiveness scale was related to the human dimension of working together, helping each other.

2.4 Results

In this research, classroom observations were made for one lesson each week for an academic year. It is in these lessons that students carried out investigations. Students sat in groups of three and worked in these groups when doing practical work. The science investigations they participated in were mostly tasks set out in the laboratory workbook or were teacher directed. The teachers tried to get to the class a few minutes before the lesson and organised the materials required for the lesson. This is described in detail in Chap. 3.

2.4.1 The Science Laboratory

The science teacher, John (pseudonym), took his class in a laboratory where his colleague Stella (pseudonym) taught all her classes. John's classes came to this laboratory four times each week. Stella kept the laboratory tidy and had her students' work displayed around the room. Also displayed were a periodic table, the school's code of conduct, charts that indicated what students needed to do to be able to gain an achieved, merit or excellence grade in the examination. The room was well lit and had seating for 30 students and ten workstations along the side benches. According to field notes about the physical laboratory environment:

> The laboratory has basic glassware neatly organised in trays and commonly used chemicals; acids, alkalis, iodine, indicators etc. in sets of ten stored on the shelf. All the work displayed on the walls is from Stella's junior science classes. No work on the walls is from students in the study class. When the students arrive, the class is clean and tidy but they often leave it in a mess with bits of paper on the floor and other rubbish in the sink. (Classroom observation)

2.4.2 Results of the SLEI

The results of the actual and preferred data from the SLEI are presented in Tables 2.3 and 2.4. Results show a just significant ($p < 0.5$) difference between the actual and preferred option for open-endedness. The open-endedness scale is particularly relevant as it measured student preference for the science investigation they carried out. The actual score (3.14) was higher than the preferred score (2.98), which indicated that they did not want the laboratory environment to be more open-ended. It can be concluded that they wanted less choice in deciding what to investigate and more teacher direction. Classroom observations during twenty lessons showed that the pattern was for the teacher to "tell" the class what investigation they were going to do. Students in the study class became comfortable with having less choice and more teacher direction which is reflected in the results (See Chap. 3).

The material environment actual score (2.81) was significantly lower ($p < 0.001$) than the preferred score (3.62) and is indicative of their preference for more equipment that worked, a less crowded laboratory, and a comfortable and attractive place to work. The science lessons for this class were held in a laboratory where the students came for only their science lessons. Their teacher was not in-charge of this laboratory. The work displayed on the walls was not their work and there was a lack of a sense of belongingness in a laboratory that felt cramped and that had with equipment that did not always work which is reflected in the results for the material environment.

Table 2.3 Paired sample statistical results for science laboratory environment inventory

Environmental factors	Mean	Standard deviation
Pair 1 Open-endedness (Actual)	3.14	0.24
Open-endedness (Preferred)	2.98	0.21
Pair 2 Material environment (Actual)	2.81	0.42
Material environment (Preferred)	3.62	0.70
Pair 3 Integration (Actual)	3.35	0.65
Integration (Preferred)	3.83	0.76
Pair 4 Rule clarity (Actual)	3.37	0.40
Rule clarity (Preferred)	3.40	0.77
Pair 5 Student cohesiveness (Actual)	2.70	0.91
Student cohesiveness (Preferred)	3.88	1.18

Table 2.4 Results of paired sample t test

Environmental factors	t	df	Sig. (2-tailed)
Pair 1 Open-endedness (Actual)—(Preferred)	2.395	16	0.03
Pair 2 Material environment (Actual)—Preferred)	−4.573	17	0.00
Pair 3 Integration (Actual)—(Preferred)	−1.969	20	0.06
Pair 4 Rule clarity (Actual)—(Preferred)	−0.160	19	0.86
Pair 5 Student cohesiveness (Actual)—(Preferred)	−4.308	21	0.00

On the cohesiveness scale, the environment was significantly ($p < 0.001$) less cohesive (2.70) than students would have preferred (3.88), indicating that they would have liked to get along with each other and preferred to be able to help each other. Although students sometimes worked in groups, this was mostly because there were ten sets of equipment and the corresponding number of workstations. Students were not required to work collaboratively to plan and carry out their investigations or to support each other's learning.

Rule clarity was high (3.37) and this was preferred (3.40), demonstrating students had, or wanted to have, a clear understanding of the classroom rules and guidelines. Integration was seen to be high (3.35) and it was preferred (3.83), indicative of the close relationship between theory and practical integration in their class. The teaching approach was to teach the science concepts first and for students to then carry out practical work to confirm the theory.

2.5 Student Experiences of Learning to Investigate

During the year students carried out a number of science investigations (Table 2.5). In the first half year, most of these investigations took two to three 50 min lessons. The first lesson was for planning and the following lesson for gathering data. It was observed that when students were gathering the data, often no time was left to find out what they had learnt. Students left the class with the instruction to complete writing the investigation for homework but inspection of their books showed that very few students did.

Table 2.5 Science investigations carried out by the study class

1	Investigating metals *Effect of acids on metals*	8	Watch that car go *Effect of slope on distance travelled*
2	Separating mixtures *Decantation, filtration, evaporation, distillation*	9	Dry ice exploration *Physical and chemical properties*
3	Energy exploration *Energy transformations*	10	Fractional distillation *Teacher demonstration of fractional distillation*
4	Body power *Work, energy and power to climb up the staircase*	11	Energy released by fuels *Burning fuels and comparing energy released*
5	Rolling marbles *Distance travelled on different surfaces*	12	DNA extraction *Extracting Cauliflower DNA*
6	Heat retention *Heat retention in paper, metal and paper cups*	13	Electrical circuits *Building simple circuits and measuring current*
7	Culturing micro-organisms *Growing bacteria and fungi on agar plates*	14	Astronomy models *Models to investigate day and night, and seasons*

The data reported here are from student responses to a questionnaire about learning and focus group interviews with five/six students three times during the year. As the researcher was accepted as a member of the class, students often came and sat down and talked to her. These informal classroom conversations were insightful in providing rich data from the willing participants. It is note-worthy that the examples cited in this book are not cherry-picked but illustrative examples contributions that illuminated something important said by the students. The first example below highlights the importance of variety in the lesson and of the student belief that it is easier to understand something when they see it work. It was also used by the teacher to confirm the theory they had been learning about energy.

2.5.1 Illustrative Example 1

2.5.1.1 Selected Practical Activity: Exploring Energy Changes

The teacher arrived early and set up in the laboratory 11 stations with different activities of energy transformation and put instructions at each station for students to follow. Students were to do the 11 activities that were set out and record their observations in a table in their workbook. They also had to identify the energy change that took place at each station. The stations included a steady stream of water to turn a wheel, solar panels to convert light energy into electrical energy, a mouse trap to move a toy vehicle, and the use of wind from a hair dryer to move a ball, to name a few.

The students listened to the instructions and chatted at the start of the lesson but once they started on the task they moved from one station/activity to the next. Of the eight groups (three students in each), seven groups were engaged and on task throughout the lesson. These 21 students appeared interested, talked to each other and asked questions. In the last quarter of the lesson the teacher asked them to share their observations and conclusions with other members of their group. The eighth group had four students—Harry, Henry, Ken, and Jake—who did not have their books. Two of them made an effort to get a photocopied sheet from the teacher but did not do any activity. For the other students, the on-task chatter and the manner in which the students carried out the tasks demonstrated that they were interested in the activities. The following information about learning is from an audio-taped conversation with students while the researcher moved around the laboratory. When students were asked about their work they said:

Pip: This is fun. I got to do the practical myself.
Jessica: It is fun because we did not have to do just one thing for the whole hour. There were lots of different things we could do.
Bob: Yeah there was variety.
Simon: I can remember the science …. (the teacher) tells us when I can see it work before my eyes.

Ed: We did not have to do heaps of writing, that's so boring. (Transcript for observed lesson)

Of the nine students asked, there were seven who were able to identify energy changes accurately. For example:

Researcher: What is this activity about?
Bob: The solar cells change light energy into electrical energy.
Researcher: What have you learnt from doing these activities?
Mili: I have learnt all sorts of energy changes like a mouse trap can change elastic potential energy into kinetic and sound.
Jessica: Light bulbs change electrical energy into light and heat and heaters change electrical energy into mostly heat but some light. (Transcript for observed lesson)

Two students who were not engaged were asked why they had not done the activity, and what they had learnt. One said "not much really", while the other, Ed, said:

Ed: Because I already know the answers.
Researcher: Can you tell me what energy changes are taking place when water is dropped on the wheel?
Ed: Gravitational potential energy into kinetic energy, just like in the hydro dams. This is dumb, we did it in year 9. (Transcript for observed lesson)

An interesting aspect was that during a focus group interview nearly six months later, students talked about the investigations they had done during the year, and all remembered energy exploration investigation as one they had enjoyed the most, and were able to say what they had learnt from it. Focus group students were able to demonstrate their understanding of the energy changes. For example:

Pip: When we are using the power pack the energy change is mostly from electrical potential to light and heat. But when we use batteries it is from chemical potential to light and heat.
Ben: And we could say that it is from chemical potential to kinetic and then to light and heat if we consider that electrons are moving.

Similarly, there was discussion about what happens when a tuning fork is hit on a mallet.

Jake: …kinetic is changing to sound, but we used the chemical potential energy stored in our cells to hit the tuning fork on the mallet. So it would be right to say chemical potential changed to kinetic which changed to light and heat.

However, the same group of students did not demonstrate an understanding of the nature of science investigation, even though they had done several fair testing type of investigation. When asked:

Researcher: Why are you taking the readings ten times?
Mili: It makes our results more valid and reliable.

Upon probing about what Mili meant about 'valid and reliable', it became apparent that none of the students understood what reliability was; it appeared to be a rote learnt answer. The point had been stressed by the teacher several times during the observed lessons as something they needed to *write* to get a better grade in the assessment.

Students completed a questionnaire about their science learning in the first school term. The question focused on their views about learning science ideas through investigating. Eighteen out of 22 students responded to this question. One student said they found it very helpful, eight found it mostly helpful, nine said it was somewhat helpful, five did not find it helpful, and four did not respond. Their responses showed that most students found this investigation helpful to some degree in their learning of science. Eight of the 18 students provided comments. They indicated that it helped them to understand the formulae they needed and how to write an equation. Others provided general comments such as: helped me to learn; made it easy because when they got to "do" the investigation it helped them to remember; and "because it lets me see the concept rather than having to 'see' it in my head" (Respondent 8); and because investigation "proves the ideas and actually show them" (Respondent 13). All five of the students who indicated investigation was unhelpful gave comments which included: "it was stupid, if a ramp is steep something will go faster" (Respondent 17); "the class mucks around" (2); "they learnt more out of the book" (1); and "I already knew this from year 9" (Respondent 3).

2.5.2   Illustrative Example 2

2.5.2.1   Heat Retention in Cups of Different Materials

This investigation was selected to study in depth because it was the investigation to prepare the students for the final assessment. The teacher wanted to familiarise students with the process they would have to follow. It took three lessons just as the assessment would. In the first lesson, students planned their investigation, in the next lesson, they carried out the investigation and collected data, and on the following day, in the third lesson, they wrote their report.

The task in the first of three lessons was to design a fair test on the insulating properties of different materials as described in the workbook (Abbott et al. 2005). In this lesson, the teacher wanted students to understand the language used in assessment tasks by asking the following questions:

What is the dependent variable here?
What is the independent variable?
Which ones would you need to control?

> Between four and five groups engaged with the investigation they were to do. The rest talked, wandered, and were not engaged. At no time during this lesson were all students on task. Ed moved to sit next to me. (Observation notes)

Researcher: Where is your plan?

Ed points to a drawing in his book.

Researcher: What are these containers?

Ed labels them.

Researcher points to the stick in each container, what is this?

Ed: Thermometers, labels them.

glass cup   metal cup   paper cup   thermometer

Researcher: What is in the containers?

Ed: Water.

Researcher: What else can you tell me about the water?

Ed: There is 100ml at 100°C in each container, and he labelled his diagram.

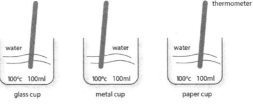

glass cup   metal cup   paper cup   thermometer

Researcher: Where will you record the temperature?

Ed: In a table.

He drew an appropriate table.

Researcher: Ed, you could have written down a complete plan and started the investigation.

Ed: Miss why would I want to do that? We did this last year and I already know the answer.

Fig. 2.1   Ed's plan for the investigation

The vignette in Fig. 2.1 is the conversation that the researcher had with Ed.

The message was clear, you learn something you did not know. Other students in the class were chatty and only a few got on with the task. Mili had collected all her data.

Researcher:	What did you learn from this investigation?
Mili:	That metal is the best, the tin.
Researcher:	Is the best what?
Mili:	It's an insulator.
Researcher:	What else did you learn?
Mili:	That the water cools fastest in the metal cup.

Evidently, Mili did not understand that if metal was an insulator the water would take longer to cool down. Later, this investigation was discussed in the focus group when several students gave their reasons for not engaging.

According to the focus group students, they already knew the science concepts because they had learnt them before. Their responses indicated that they had not considered that this investigation was to prepare them for their final assessment. The teacher's purpose for doing this investigation was to give students an opportunity to practise an investigation and learn the necessary terminology to become familiar with the process that they would need to follow in the assessment.

When students in the focus group were asked how this investigation helped them to prepare for the final assessment, the following discussion took place.

Bob: It showed us the lines which we had to follow to write out the method. How to use our time?

His response helped the others to remember doing this investigation:

Pip:	And what we were looking for when we were doing it.
Researcher:	Do you remember your teacher marked and returned your plan? Did anyone use that feedback that the teacher gave you?
Bob:	No.
Researcher:	Why?
Jake answered instead:	I tried to, but I didn't improve my mark, but I'm still happy because I passed. (Focus group interview)

Through this investigation the teacher supported students' learning by giving them time to think what they had to do for their assessment. What the teacher did not know was that these students had already done this investigation the previous year.

2.6 Chapter Summary

The laboratory holds an important place in secondary schools, and classroom observations showed that most students, most of the time, were involved in "doing" practical work. In almost all cases, the investigation they were to do was either set out in the workbook or students were told what they were going to do. It is not surprising that students did not want open-ended investigations as it was easier to do what "they were told" and it then became a habit. This is contrary to Rudduck and McIntyre's (2007) research where students wanted autonomy and control over their learning. As can be expected, students wanted to have equipment easily available and in working order. Sometimes not having all the equipment was frustrating for them. They said that they wanted to work with their peers, and at times they worked in groups. However, the students wanted to sit with their friends but this was discouraged by the teacher who believed that it would lead to off-task behaviour. The common pedagogical approach was to teach the science ideas first and then to do the practical work to confirm what they had learnt in the theory lesson. This was problematic in that students were developing the thinking that by following a set of steps sequentially, they could get to the "correct" answer in order to achieve a better grade.

Ahmad et al. (2013) found a predictive relationship between teachers' perception of the learning environment and the satisfaction they experienced in their teaching. The results of the SLEI were useful as they provided the teacher with insight into students' perceptions about their current laboratory environment and how they would prefer it to be. Luketic and Dolan (2013) argue that teachers can respond to the SLEI results and make appropriate changes to make the laboratory environment conducive to student learning. These results were shared with the teacher who had the option to take the findings on board and consider how he could modify his pedagogical approach to address the learning needs of his students.

It is noteworthy that students enjoyed, and appeared to have learnt most from, exploration of a series of short practical activities set up as work stations. They explained their enjoyment in terms of the variety offered as opposed to doing one investigation for the whole hour. Another reason offered for enjoyment was doing the activity and not having to write too much about it. It was interesting that students in the focus group collectively could remember a number of investigations that they had carried out and what they had learnt from them. However, they did not appear to distinguish between 'investigations', whether it be fair testing or another type, and other practical work.

From the teacher's perspective, it was important to give the students an opportunity to learn the process they would need to follow for the final assessment, but this had not been communicated to the students. The teacher did not know that the students had in the past done the heat investigation. So, for those who understood the teacher's intention for learning, learnt from it. For those who thought it was about the science ideas that were to emerge from doing it, perceiving that they already knew them, there was no learning. It was indeed important for the teacher and the students to find out what they had learnt, but no time remained in the lesson for reflection.

References

Abbott, G., Cooper, G., & Hume, A. (2005). *Year 11 science (NCEA level 1) workbook.* Hamilton, New Zealand: ABA.

Abrahams, I. (2009). Does practical work really motivate? A study of the affective value of practical work in secondary school science. *International Journal of Science Education, 31*(17), 2335–2353.

Abrahams, I., & Millar, R. (2008). Does practical work really work? A study of the effectiveness of practical work as a teaching and learning method in school science. *International Journal of Science Education, 30*(14), 1945–1969.

Ahmad, C. N. C., Osman, K., & Halim, L. (2013). Physical and psychosocial aspects of the learning environment in the science laboratory and their relationship to teacher satisfaction. *Learning Environments Research, 16*(3), 367–385.

Dkeidek, I., Mamlok-Naaman, R., & Hofstein, A. (2012). Assessment of the laboratory learning environment in an inquiry-oriented chemistry laboratory in Arab and Jewish high schools in Israel. *Learning Environments Research, 15*, 141–169. doi:10.1007/s10984-012-9109-3.

Fraser, B. J., & McRobbie, C. J. (1995). Science laboratory classroom environments at schools and universities: A cross-national study. *Educational Research and Evaluation, 1*(4), 289–317.

Fraser, B. J., McRobbie, C. J., & Giddings, G. J. (1995). Development and cross-national validation of a laboratory classroom environment instrument for senior high school science. *Science Education, 77*, 1–24. doi:10.1002/sce.3730770102.

Hodson, D. (1990). A critical look at practical work in school science. *School Science Review, 70*(256), 33–40.

Hodson, D. (1991). Practical working science: Time for a reappraisal. *Studies in Science Education, 19*, 175–184.

Hofstein, A. (2004). The laboratory in chemistry education: Thirty years of experience with developments, implementation and research. *Chemistry Education Research and Practice, 5*(3), 247–264.

Hofstein, A., & Lunetta, V. N. (1982). The role of the laboratory in science teaching: Neglected aspects of research. *Review of Educational Research, 52*(2), 201–217.

Hofstein, A., Kipnis, M., & Kind, P. (2008). Learning in and from science laboratories: Enhancing students' meta-cognition and argumentation skills. In C. L. Petroselli (Ed.), *Science education issues and developments* (pp. 59–94). London: Nova Science.

Hofstein, A., Levy-Nahum, T., & Shore, R. (2001). Assessment of the learning environment of inquiry-type laboratories in high school chemistry. *Learning Environments Research, 4*, 193–207.

Luketic, C. D., & Dolan, E. L. (2013). Factors influencing student perceptions of high-school science laboratory environments. *Learning Environments Research, 16*, 37–47. doi:10.1007/s10984-012-9107-5.

Millar, R. (2010). Practical work. In J. Osborne & J. Dillon (Eds.), *Good practice in science teaching: What research has to say* (2nd ed.). Maidenhead: Open University Press.

Osborne, J. (1998). Science education without a laboratory? In J. Wellington (Ed.), *Practical work in school science: Which way now?* (pp. 156–173). London: Routledge.

Rudduck, J., & McIntyre, D. (2007). *Improving learning through consulting pupils.* Abingdon: Routledge.

Toplis, R. (2012). Students' views about secondary school science lessons: The role of practical work. *Research in Science Education, 42*, 531–549. doi:10.1007/s11165-011-9209-6.

Woolnough, B. E. (1991). Setting the scene. In B. E. Woolnough (Ed.), *Practical science* (pp. 3–9). Milton Keynes: Open University Press.

Chapter 3
Motivation to Learn Science Investigation

3.1 Relevant Theories of Motivation

Investigative work in science is seen to provide opportunities for students to engage with, and influence, their own learning (Toplis 2012). Wellington (1998, 2005) drew attention to a lack of research in the affective domain of learning science investigation. In the context of learning through investigation, Palmer (2005) argued that if learning requires effort, then motivation is also required because students are unlikely to make the effort unless they are motivated. Lin et al. (2013) investigated students' scientific epistemological beliefs and motivation to learn science and found that in China and Taiwan there was a link between cultural background and motivation to learn science. A number of theories have been put forward to explain and help predict behaviour (McInerney and McInerney 2002). Some relevant motivational theories have been reviewed here.

Brophy (1987) posits that "motivation to learn is an enduring disposition to value learning as a worthwhile and satisfying activity, and thus to strive for knowledge and mastery in learning situations" (p. 200). It leads to task engagement and effort to acquire knowledge or to master a skill. According to constructivist theory, learning is viewed as an active process that requires effort on the part of the learner (Driver 1989).

Palmer (2009) contends that motivation is needed to initiate and maintain learning and is central to all action, and that motivation is an essential "pre-requisite and co-requisite" for learning (p. 147).

The role of motivation during learning to investigate is pivotal (Schunk 1991); students who are motivated to learn are likely to participate in activities that they think will help them to learn (Meece 1991). They will engage in learning behaviours such as taking notes, reading the material over and over, be focused when instructions are given, and making sure they understand and asking for help when they do not. In contrast, students who are not motivated to learn have the opposite traits, and as a consequence their learning suffers.

© The Author(s) 2015
A. Moeed, *Science Investigation*, SpringerBriefs in Education,
DOI 10.1007/978-981-287-384-2_3

Pintrich and Schunk (2002) emphasise "that motivation bears a reciprocal relation to learning and performance; and what students do and learn influences their motivation" (p. 6). When students achieve learning goals, goal attainment conveys to them that they possess the requisite capabilities for learning. These beliefs motivate them to set new, challenging goals. Students who are motivated to learn often find that once they are, they are intrinsically motivated to continue their learning (Meece 1991). These are the students who are most likely to want to improve their performance and achieve.

In the 1960s, reinforcement theory dominated the educational literature. This theory conceptualises motivation entirely in terms of observable behaviour. In the educational context, it considers motivation not as a trait but as a set of behaviours and whether the individual is rewarded or punished, believing motivation to be extrinsic. Behaviour is reinforced through external positive reinforcement (reward) or negative reinforcement (punishment). Behavioural theorists contend that explanations for motivation do not need to include thoughts and feelings but that people are motivated by environmental events (Pintrich and Schunk 2002). From reinforcement theory arose the cognitive motivation theories.

Cognitive theories emphasise learners' thoughts, beliefs, and emotions. Proponents of these theories found reinforcement theory and the assumptions made about behaviour unsatisfactory and started to explore psychological variables that cannot be observed directly. Cognitive theories of learning suggest that the key to people's motivation is the desire to solve problems, have insight, and gain understanding, particularly in ambiguous or problematic situations (McInerney and McInerney 2002). An essential element in a cognitive perspective of motivation is the concept of intrinsic motivation. According to Ryan and Deci (2000):

> Intrinsic motivation.... refers to doing something because it is inherently interesting or enjoyable, and extrinsic motivation...refers to doing something because it has a separable outcome. (p. 55)

Intrinsic motivation generates "instinctive pleasure" when the learner succeeds in learning something new or completes a challenging task. Successful completion of a task leads to confidence and the learner is more likely to engage in learning activities as a consequence. "Intrinsic motivation is characterised by enthusiastic task involvement, a want to experience novelty and adventure, striving for excellence, trying to understand something, wanting to improve and seeing a purpose in doing the task" (Ryan and Deci 2000, p. 208). Intrinsically motivated behaviours are undertaken out of interest. McInerney and McInerney (2002) highlight the nexus between intrinsic motivation and cognitive theories of learning as: "a facility for learning that sustains the desire to learn through the development of particular cognitive skills" (p. 208). In the context of learning to investigate, it can be argued that if a student has the choice of a question they are interested in investigating, it is more likely that they will engage in the process, they might make a real effort to find the answer to it, develop an understanding, and perhaps even strive for excellence. In contrast, extrinsic motivation depends on external factors such as rewards, attention and praise. Deci et al. (2002) questioned its effectiveness for learning in the long term.

Cognitive motivation theories include attribution and goal orientation theories. In attribution theory an explanation is given for certain behaviour. For example, a student gets a low mark in a test. They may attribute the low mark to the test having questions that were on material not taught by the teacher. Alternatively, a student may think the reason they performed poorly was due to their own lack of ability. In other words, attribution theory describes the processes of explaining events and the behavioural and emotional consequences of those explanations.

Weiner (1986) posits that a motivated high achiever will approach rather than avoid tasks related to succeeding because they believe success is due to high ability and effort and failure is caused by bad luck or a poor examination, i.e., not their fault. Their self-esteem is not hurt by failure but when they are successful it builds their confidence. These people persist with a task rather than give up because they assume that failure is caused by lack of effort, something they can change. They select moderately difficult tasks because succeeding in these tells them how well they are doing. These individuals are energetic learners as they believe that success is proportional to hard work. Unmotivated people, on the other hand, avoid success-related tasks as they doubt their own ability and assume success is related to factors beyond their control and, as a consequence, even when they do succeed they do not feel responsible for their success, thus success does not increase their pride. They give up because they attribute failure to lack of ability. They do not believe that success is related to effort, so they lack enthusiasm and drive. Avoidance strategies used by these learners include procrastination, making excuses, avoiding challenging tasks, and not trying (Eccles and Wigfield 2002). Weiner (1992) posits that how an individual interprets their achievement outcome is more important than either their motivational disposition or actual outcome and is an influential factor that determines their future quest for achievement. Weiner believes that ability, effort, task difficulty and luck are most important achievement attributions.

In their review of motivational beliefs, values, and goals, Eccles and Wigfield (2002) classified related theories of motivation into five major categories. These are:

> Theories focused on expectancies for success (self-efficacy theory and control theory); theories focused on task value (theories focused on intrinsic motivation, self-determination, interest and goals); theories that integrate expectancies and values (attribution theory, the expectancy value model, and self-worth theory); theories integrating motivation and cognition (social cognitive theories of self-regulation and motivation); and theories of motivation and volition. (p. 109)

Modern expectancy-value theory is based on an expectancy-value model that links achievement, performance, persistence, and choice directly to individuals' expectancy related and task value beliefs (Wigfield and Eccles 2000). In self-worth theory, Covington (1992) defines the motive for self-worth as the tendency to establish and maintain a positive self-image. Learners need to believe that they are academically competent in order to think they have self-worth. In the school context, however, school evaluation, competition, and social comparison make it difficult for many learners to maintain the belief that they are competent academically. These learners develop strategies to avoid appearing to lack ability.

Goal theories link achievement goals and achievement behaviour. Goal theorists have used a number of terminologies to conceptualise goal orientation. According to Covington (1998), there are three goal orientations: (1) learners with ego-involved goals want to out-perform others; (2) mastery oriented learners choose challenging tasks and focus on their own progress and are not concerned about how others perform; (3) learners with performance goals take on tasks that they know they can do. Their short-term goal is to complete a task. According to Covington (1998), performance goals can be further classified into performance-approach goals where engagement with a task is for performance reasons. Conversely, learners with performance avoidance goals are disengaged so that they are not considered incompetent. Whatever their goal orientation, to achieve, learners need to engage with a task, and be willing to carry it out; they need to get started and then continue.

Motivation to learn requires volition to carry out the learning and it has two elements—"the strength of will" needed to complete the task and the "diligence of pursuit" (Corno 1992, p. 14). Corno and Kanfer (1993) argue that motivational processes only lead to the decision to act. Once the individual engages in action, volitional processes take over and determine whether or not the intention is fulfilled.

Eccles and Wigfield (2002) posit that secondary school students need autonomy and self-control. If the school structures do not allow for this, it results in a mismatch between the students' needs and whether they can be satisfied by opportunities and choices offered by the school. If this is not so, the students' motivation to learn is affected (Meyer et al. 2006).

Pintrich and Schunk (1996) suggest interest as an effective motivator and, further, that interest is related to increased memory, greater comprehension, deeper cognitive engagement, and thinking. Rennie et al. (2001) have found that Australian adolescents view science as a dull and boring subject that fails to motivate them. More recently, Palmer (2009) reported that many science students are lacking in motivation and it has been argued that without motivation little learning occurs. Therefore, it is essential for science education to deal with the issue of student motivation. Perhaps one way to address this in science and science investigation is through creating situational interest. Features that arouse situational interest include personal relevance, novelty, activity level, and comprehension (Eccles and Wigfield 2002).

Palmer (2009), in his study of situational interest during an inquiry science lesson, found that students identified learning as the most common reason for their interest in the investigation. He says that this is possibly because when "one learns something, one is learning something new" (p. 160). Palmer considers situational interest as short-term interest that is generated by particular situations. Examples of such interest are experiments that have a 'wow' factor that can provide short-term interest for students who might not otherwise be interested. Other factors influencing interest are choice but he asserts it has to be "meaningful choice", and "cognitive engagement". Palmer found hands-on activity that involves moving from seats, offers variety and had a social dimension of working with others is likely to lead to student engagement. He argues that the strongest reason for interest is perhaps the novel approach to learning for the students because he believes situational interest is very powerful and overrides any other motivational orientation the students may have.

3.2 Assessment of Motivation

With respect to assessment of motivation, commonly employed indices are choice of task, effort, persistence, and achievement (Pintrich and Schunk 2002). While there is disagreement among the researchers about the nature of motivation and how motivational processes operate, there is some agreement that behavioural indicators can be used to determine the presence or absence of motivation. Brophy (1987) argues that despite the intuitive appeal, choice of task is not a useful index of motivation in schools because in many classrooms the students typically have few, if any, choices. Brophy adds that students who are motivated to learn are willing to put effort into learning. This learning could be of skills that require physical or cognitive effort. Effort, therefore, is an appropriate index. Those students who are motivated to learn could spend time on the task and will continue to do so even if they come up against obstacles.

There are three ways used to assess motivation—direct observations, ratings by others, and self-reports (Zusho and Pintrich 2003). Direct observations would include looking for choice of task, effort expended, and persistence. However, one can only focus on overt actions and therefore these can be superficial and may not capture the essence of motivation. Direct observations also ignore the cognitive and affective processes underlying motivational behaviour. Rating by others could include the parents or teachers giving a rating on a scale, giving data that cannot be attained from direct observations. On the other hand, it involves the person providing the rating to remember what they have observed. As memory is selective and constructive, ratings may not be valid indicators but could be useful in conjunction with observations. Self-reports involve people's judgements and statements about themselves. Ways of self-reporting include questionnaires, interviews, stimulated recalls, and dialogues. In this research, a number of these methods have been used to assess student motivation to learn to investigate, including observations, student interviews, incidental conversations, and questionnaires.

3.3 Motivation and Approaches to Learning

Motivational theorists (Entwistle 2005; Entwistle and Ramsden 1983) describe the relationship between student motivation to learn and their approach to learning. The learner can have a deep or surface approach to learning which is dependent upon what they want and how they want to learn, as well as the nature of assessment of this learning. Entwistle (2005) defines the features of deep and surface approaches to learning as:

Deep Approach
Intention to understand material for oneself
Interacting vigorously and critically on content
Relating ideas to previous knowledge/experience
Using organizing principles to integrate ideas
Relating evidence to conclusions
Examining the logic of the argument

Surface Approach
Intension simply to reproduce content
Accepting ideas and information passively
Concentrating only on assessment requirements
Not reflecting on purpose or strategies in learning
Memorising facts and procedures routinely
Failing to recognize guiding principles or patterns. (p. 3)

A deep approach is linked to academic interest in a subject along with self-confidence, and a surface approach is linked to anxiety and fear of failure. Mastery goals have been found to be most beneficial and are related to deep learning as opposed to performance goals that lead to a surface learning approach (Palmer 2005).

3.4 Motivational Framework for the Study

The selection of a framework for conceptualising motivation requires that elements of several theories of motivation should be considered. For this research, goal theory is one appropriate lens as the study investigates students' learning of investigation, focussing on assessment. Attribution theory would be useful to compare what in students' views are reasons for their success or otherwise when carrying out an investigation. Expectancy and value theories may play a role in determining what leads to achievement, what makes students persist and perform, and whether having choice influences students' motivation to learn science investigation. In understanding whether students are willing to learn to investigate and what keeps them going may be found in volition theory. As is evident in recent research, situational interest may have a role to play.

3.5 Results

In this section, results related to task engagement, variety, and novelty are reported. As with learning, the data were collected through classroom observations, incidental conversations with the students during the lesson, a questionnaire, and focus group interviews.

3.5.1 Engagement

Student engagement is considered to be an indicator of motivation by most teachers (Moeed 2010). Table 3.1 shows a summary of student engagement during each of the observed lessons in school term 1. The data were collected on the observation schedule. In each quarter of the lesson, student engagement was recorded by counting the number of groups on task. This information was then compared with the running record to determine what was happening in the class at the time. An iterative process was used to check the accuracy of the recording. The number with a negative sign indicates the number of groups that were not engaged during that period.

By the end of the term, based on the observations and records, it can be said that there were some students who were attentive and picked up what was needed by reading instructions or listening to the teacher. As the term progressed, students were less attentive at the start of the lesson and less than half were engaged in the last quarter of the lesson. Sometimes, their chatter made it difficult for the teacher to get the instructions across and some of the students at the back who did want to listen were not able to do so.

Table 3.2 shows student engagement during each observed lesson in term 2. The SLEI was administered in observed lesson nine. The students did both forms of the SLEI, "actual" followed by "preferred", while the teacher was on leave. The inventory was administered by the researcher and most students completed each version in about 20 min. Engagement during this inventory is reported in lesson 9 and the results in Chap. 2.

Towards the end of the term the researcher observed that:

> The noticeable difference is that students are taking a lot longer to settle down at the start of the lesson. The teacher has decided to continue to talk over their noise to give instructions. (Observation notes)

Table 3.1 Student engagement in year 11 science class in term 1

Student engagement in each quarter	Student engagement in lessons 1–5				
	1	2	3	4	5
	Investigating metals	Separating mixtures	Feedback on test	Energy exploration	Body power
	16 March	23 March	30 March	6 April	13 April
First ¼	All	Most	None	All	None
Second ¼	All	Less than half −3 groups	Less than half	Most −1 group	Half class
Third ¼	Most −2 groups	Most (during practical)	Less than half 4 groups listening	All	All during practical
Fourth ¼	Most −2 groups	Most	Less than half	Half	None

Note No data are reported about groups 6 and 8 as these groups had students who did not participate in this research

Table 3.2 Student engagement in year 11 science class in term 2

Student engagement in each quarter	Student engagement in lessons 6–10					
	6	7	8	9	10	
	Rolling marbles	Planning heat retention (formative)	Data gathering heat retention (formative)	Science laboratory environment inventory	Culturing micro-organisms	AS1.1 watch that car go
	4 May	11 May	18 May	25 May	1 June	6 June
First ¼	None	Half −groups 4 and 5	Few −groups 3, 4 and 5	All	None	All
Second ¼	Half class	Half −groups 4 and 5	Few −groups 4 and 5	All	Half class	All
Third ¼	All during practical	Few	Few −groups 4 and 5	Most	All during practical	All
Fourth ¼	None	Few	Few most finished and very poorly behaved	Few after completing the inventory	None	Most finished and restless

Note No data are reported about groups 6 and 8 as these groups had students who did not participate in this research. The groups identified in the table with a negative sign indicate the groups in class that were not engaged during that period of the lesson

The teacher continued to do practical work and investigation through terms 3 and 4. The summary of student engagement in the class is presented in Table 3.3.

At the start of term 3, students engaged with the work in the first week and worked for almost the entire lesson but from the following week a trend appeared where students were less and less engaged as each lesson progressed. Most lessons ended with students leaving without tidying up.

Another notable point was that in each observed lesson the students had undertaken practical activity but the lessons ended without summing up and finding out what they had learnt. It would appear that the engagement had only been about doing an enjoyable activity. The focus group students pointed out that it was a preferred alternative to writing.

3.5.2 Variety

Task variety generated interest for students when they explored energy changes in term 1 and became enthusiastically involved. They commented that they liked variety rather than doing the same task for the whole hour (see Sect. 2.5.1).

3.5.2.1 Illustrative Example: Exploration of Dry Ice

This activity was selected as an illustrative activity because in year 9 students had done similar exploration of dry ice and it demonstrated that those who

Table 3.3 Student engagement in year 11 science class in terms 3 and 4

Student engagement in each quarter	Student engagement in lessons							
	12	13	14	15	16	17	18	19
	Jigsaw	Dry ice	Fractional distillation	Energy released by fuels	Extracting cauliflower DNA	Mock exam review	Electrical circuits	Astronomy models
	20 July	3 August	10 August	17 August	31 August	12 October	19 October	26 October
First ¼	Most –group 5	Most –group 5	Some– groups 3, 4, 5	Few	Most –group 5	Few	Some –groups 3, 5, 7	None
Second ¼	All	All	Few	Few	Few	Few	Some –groups 3, 5, 7	Few
Third ¼	Most –group 5	Few	Few	None	Few	Few	Some –groups 3, 5, 7	Few
Fourth ¼	All	None	None	None	Some –groups 3, 4, 5, 7	Some –groups 3, 4, 5, 7	Some –groups 3, 5, 7	Most –group 5

remembered the science ideas either did not engage or treated it as play. The activity was set up with stations around the laboratory like the energy exploration the class had done in term 1. Students were told that they would be doing dry ice activities if they paid attention in the first part of the lesson. The teacher gave them instructions after doing a recall quiz for the first 10 min of the lesson. The students moved around the room and followed the instructions written on the cards placed next to the equipment.

> There is a lot of excitement and students are going around and doing the activities. The noise level is getting higher and higher. Bob and Sarah are not engaged which is unusual. Sarah is quietly reading a fiction book. (Observation notes)

Bob came and sat down next to the researcher, which was an opportunity to find out what he had learnt and why he was not doing the activities:

Researcher: What did you learn?
Bob: How to be an idiot and muck around. (Points to those being silly)
Researcher: What did you learn about dry ice?
Bob: I learnt nothing new.
Researcher: Can you tell me a science idea for each of these activities?
Bob: Dry ice changes from solid to gas, does not turn to liquid. It sublimates. Carbon dioxide is heavier than air so it can blanket a fire and put it out. So we use it in fire extinguishers.
Researcher: So carbon dioxide puts fires out by having a chemical reaction with the fire?
Bob: No it cuts off oxygen by making a blanket over the fire. Fuel needs oxygen to burn.
Researcher: What else did you learn about dry ice?
Bob: When it dissolves in water it makes it acidic. Blue litmus turns red. That rude noise it makes is because when you press down with spoon it is sublimating and tries to escape and you're putting pressure on it. If you push it along the table top it glides like a hovercraft.
Researcher: How did you learn all this?
Bob: We learnt this last year and in year 9. (Transcript of informal audio-taped conversation)

After the above exchange, Bob went back to his seat. The rest of the class continued to play with dry ice until the end of the period. There was no time for the teacher to close the lesson and find out what they had learnt.

3.5.3 Novelty

An interesting theme emerged and indicated that students engaged with, and learnt from investigation where they had done something that they had either not thought about or experienced. Investigating fuels was one such investigation.

3.5.3.1 Illustrative Example: Measuring Energy Content of Fuels

This investigation was selected because it was a fair testing type of investigation, but for most students this was new learning. It was one of the lessons where there were glimpses of a 'wow' element. Students in the focus group remembered and talked about it so it was a memorable event for some. The purpose of the investigation was to compare the energy released by different fuels. Students were asked to use the templates in their workbook to plan this investigation and then to carry it out. They talked about dependent and independent variables. The teacher walked around the room to check.

> Students except for the two rows in the front are making no effort to get started and are wasting time. It appears they have no idea as to what needs to be done. Basically they need to put 50 ml of water in a test tube and put it on the retort stand. Measure the initial temperature of water. Then collect some fuel in a bottle top. Put it under the test tube and light it. When it stops burning they have to measure the final temperature of water. Repeat this with each fuel and compare the rise in temperature each time. (Observation notes)

The teacher went through safety reminders. Students were interested and keen to get started. There was a lot of excitement in group 3 in the front when they burnt diesel and saw bits of carbon and a lot of smoke coming out. The three groups that almost always got on with the task did the task as requested and worked out which fuel released most energy:

Researcher: What did you find out?
Susan: Miss, you can't see when alcohol and kerosene burn.
Researcher: Then how did you know they burnt?
Susan: The temperature went up. It went up a lot, alcohol was the hottest.
Researcher: The hottest?
Susan: It heated the water and it made it boil.
Researcher: And which fuel gave the least energy?
Susan: Definitely diesel, and look at the test tube it is all black and yucky.
Researcher: And why was that?
Susan: Diesel does not burn cleanly, we should not use it, it pollutes the air. (Transcript of audio-taped casual conversation)

Another student, Harry, had set up the equipment. He was about to light the fuel that was placed under an empty test tube with a thermometer in it. When asked why he was about to heat an empty test tube he said, "that is what Ken is doing". When probed further it became clear that Harry had not read the instructions or listened to the teacher. He just decided to copy what his friend was doing and had not noticed that Ken had water in his test tube.

All six participants came to the last focus group interview. They were asked questions about their learning, science experiences outside the classroom, who they asked for help, what they enjoyed about science, and if they had the freedom what they would like to investigate. First, they were asked to give some examples of science ideas that they had learnt through investigating during the year:

Jake: Probably one where we burnt different fuels, different products of crude oil like octane and kerosene. And people rant on about how certain chemicals like diesel are more toxic because they produce more fumes but you don't get to see it that much. But when you actually see smoke and soot, what actually happens when it burns, like the difference between octane and kerosene or diesel and that, you can actually see, or in some cases smell the difference. (Jake)

Researcher: Which one of those gave out more smoke and smell?

Bob: Oh definitely kerosene, it's heavier.

Ed: I had fun catching the carbon that was floating in the air.

Pip: My favourite was probably the electrical one about parallel circuits, the way the current divided and completed a circuit. You don't really know if that's true or not but then when you do it you see how the light changes and all that.

Jeff: Yeah I know that, because when they tell you you're not too sure if it actually does but when you get to investigate you can actually see it happen before your eyes. (Third focus group interview)

The others could also name their favourite investigation and what they had learnt from it. Next they were asked what topic in science they enjoyed the most and why? There was agreement that they all liked chemistry best.

When asked about what other science they had done that they had really enjoyed, Jeff, Bob, and Jessica remembered visiting Victoria University of Wellington:

Bob: It was free. It was great because we learnt about radioactive materials and we actually got to see first-hand. They had samples of radioactive materials that produce quite a bit of radiation and we were able to measure how much radioactivity was being pumped out by all these things.

Pip: We got to use different instruments that we don't have at school. (Third focus group interview)

Students were asked what they would like to investigate if they could do anything they liked. Pip wanted to do fieldtrips. She had loved going to the rocky shore in primary school and being able to explore. Others agreed that fieldtrips were great even if all you saw was a cow (on a recent geography trip). Jessica wanted to investigate gun powder:

If I was allowed to do it, legally, you see I do shooting… so we have to load our own bullets with different gun powders for different ranges. And I was thinking well maybe since the bullet… maybe you can make different grades of gun powder and test how powerful it is, how much kick it's got behind it. That is what I would like to investigate. (Jessica, third focus group interview)

This suggests that Jessica understood about investigating; she had a question and proposed a way to answer it.

Here is what Bob, who expressed real interest in science, wanted to explore and look for evidence:

> Probably jump in a rocket and go around the solar system. Wanting to prove, because we learn about how there's all these different things out there, but you don't really know. I just want to prove, not to anyone else, just to myself that it is true and they exist. (Bob, third focus group interview)

Overall, this group of students had participated in science (to varying degrees) and they had all appeared to have learnt some science through investigating. Most were curious and knew what science they would like to do if given the opportunity.

The survey results showed that most students knew the value of doing the investigation because it would be assessed and was worth credits and grades. This is discussed in detail in the next chapter. The last words on motivation can be illustrated in the following response to survey questions:

Question: When you are in class and need help with science who do you ask for help?

Chris: I ask Nicole.

Question: When you are at home and need help with science who do you ask for help?

Chris: I ring Nicole.

Question: How do you know when you understand a science idea?

Chris: Daah! I can explain it to Nicole!

Question: Is there any thing else you want to say.

Chris: Yes, she ROCKS my socks!!

A reminder that these are 15-year-old boys and girls, and what happens in the science class may not be so motivational.

3.6 Summary

The study showed that engagement, variety and novelty of task were factors that influenced student motivation to learn science investigation. In most lessons observed, task engagement was highest when the students were *doing* investigation. For most, it meant a reprieve from written work. Most students engaged in the dry ice activities even though they had done this in the past. It offered variety, and moving around the lab with their friends was motivational. However, for one capable student who was almost always positive and engaged, this was not motivational at all. He knew the answers and considered it a waste of time. Learning something new when burning the fuels was highly engaging and students were pleasantly surprised with the finding that diesel produces soot and made the link to life where they had heard that diesel is a pollutant. This investigation also created situational interest. It is important to remember that adolescents may find other things in life motivational.

References

Brophy, J. (1987). Synthesis of research for motivating students to learn. *Educational Leadership, 45*(2), 9–17.

Corno, L. (1992). Encouraging students to take responsibility for learning and performance. *Elementary School Journal, 93*, 69–83.

Corno, L., & Kanfer, R. (1993). The role of volition in learning and performance. *Review of Research in Education, 19*, 301–341.

Covington, M. V. (1992). *Making the grade: Self-worth perspective on motivation and school reform.* New York: Cambridge University Press.

Covington, M. V. (1998). *The will to learn: A guide for motivating young people.* Cambridge University Press, Cambridge.

Deci, E. L., Koestner, R., & Ryan, R. M. (2002). Extrinsic rewards and intrinsic motivation in education: Reconsidered once again. *Review of Educational Research, 71*, 1–27.

Driver, R. (1989). Students' conceptions and the learning of science. *International Journal of Science Education, 11*(5), 481–490.

Eccles, J. S., & Wigfield, A. (2002). Motivational belief, values, and goals. *Annual Review of Psychology, 53*, 109–132.

Entwistle, N. (2005). *Learning and studying: Contrasts and influences.* Retrieved September 8, 2005 from http://www.newhorizons.org/future/Creating_the_Future/crfut_entwi9stle.html.

Entwistle, N., & Ramsden, P. (1983). *Understanding student learning.* London: Croom Helm.

Lin, T. J., Deng, F., Chai, C. S., & Tsai, C. C. (2013). High school students' scientific epistemological beliefs, motivation in learning science, and their relationships: A comparative study within the Chinese culture. *International Journal of Educational Development, 33*(1), 37–47.

McInerney, D. M., & McInerney, V. (2002). *Educational psychology: Constructing learning.* Australia: Pearson.

Meece, J. L. (1991). The classroom context and students' motivational goals. In M. L. Maehr, & P. R. Pintrich (Eds.), *Advances in motivation and achievement: A research annual: Vol. 7.* Greenwich, CT: JAI Press.

Meyer, L., McClure, J., Walkey, F., McKenzie, L., & Weir, K. (2006). *The impact of the NCEA on student motivation: Final Report to Ministry of Education.* Wellington: Ministry of Education.

Moeed, H. A. (2010). Science investigation in New Zealand secondary schools: Exploring the links between learning, motivation and internal assessment in Year 11. (Unpublished Doctoral diss). New Zealand: Victoria University of Wellington.

Palmer, D. C. (2005). A motivational view of constructivist-informed teaching. *International Journal of Science Education, 27*(15), 1853–1881.

Palmer, D. H. (2009). Students' interest generated during an inquiry skills lesson. *Journal of Research in Science Teaching, 46*(2), 147–165.

Pintrich, P. R., & Schunk, D. H. (1996). Motivation in education: Theory, research and applications. Columbus OH: Merrill Prentice Hall.

Pintrich, P. R., & Schunk, D. H. (2002). *Motivation in education: Theory, research and applications* (2nd ed.). Columbus, OH: Merrill Prentice Hall.

Rennie, L. J., Goodrum, D., & Hackling, M. (2001). Science teaching and learning in Australian schools: Results of a national study. *Research in Science Education, 31*, 455–498.

Ryan, R. M., & Deci, E. L. (2000). Self-determination theory and the facilitation of intrinsic motivation, social development, and well-being. *American Psychologist, 55*(1), 68.

Schunk, D. (1991). Self-efficacy and academic motivation. *Educational Psychologist, 26*, 299–323.

Toplis, R. (2012). Students' views about secondary school science lessons: The role of practical work. *Research in Science Education, 42*(3), 531–549.

Weiner, B. (1986). *An attributional theory of motivation and emotion.* New York: Springer.

Weiner, B. (1992). *Human motivation: Metaphor, theories and research.* Newbury Park, CA: Sage.

Wellington, J. (1998). Practical work in science: Time for re-appraisal. In J. Wellington (Ed.), *Practical work in science. Which way now?* (pp. 3–15). London: Routledge.

Wellington, J. (2005). Practical work and the affective domain: What do we know, what should we ask, and what is worth exploring further? In S. Alsop (Ed.), *Beyond cartesian dualism* (pp. 99–109). Netherlands: Springer.

Wigfield, A., & Eccles, J. S. (2000). Expectancy-value theory of achievement motivation. ContemporaryEducational Psychology, *25*, 66–81.

Zusho, A., & Pintrich, P. R. (2003). A process-oriented approach to culture: Theoretical and methodological issues in the study of culture and motivation. In F. Salili & R. Hoosain (Eds.), *Teaching, learning, and motivation in a multicultural context* (pp. 33–65). USA: Information Age.

Chapter 4
Assessment of Science Investigation

4.1 Multiple Purposes of Assessment in Science

Assessment is an essential element in teaching and learning and is used to: find out what the learner knows prior to introducing new learning (diagnostic assessment; Hackling 2005); during the learning process to ascertain and enhance learning (formative assessment; Bell 2005); and for gaining accreditation of a qualification for which students have been studying (summative assessment; Biggs 2003; Hall 2006). Summative assessment has an important role in communication between an institution and the wider community and, therefore, the results need to be trustworthy (Broadfoot and Black 2004; Jones 2007).

Assessment has the following four purposes.

4.1.1 Diagnostic Assessment

Diagnostic assessment is assessment *before* learning. It is used to ascertain prior knowledge and alternative conceptions or misconceptions the learner may have so that teaching can be planned to build on students' existing knowledge and further, to address their alternative conceptions (Hackling 2005). A key component in a constructivist approach to science teaching is to find out what the students' alternative conceptions, or preconceptions are before planning the next learning steps (Birk and Kurtz 1999; Driver et al. 1985; Osborne 1982). Diagnostic assessment is learner centred and a critical aspect of constructivist pedagogy. There was little evidence of the use of diagnostic assessment in the study class.

© The Author(s) 2015
A. Moeed, *Science Investigation*, SpringerBriefs in Education,
DOI 10.1007/978-981-287-384-2_4

4.1.2 Formative Assessment

Formative assessment is assessment *for* learning (Crooks 2004). It is about making use of informal and sometimes formal ways to find out what is being learned. It is very powerful as it provides the learner with feedback on how learning can be enhanced. This kind of assessment is learner centred, usually takes place during learning, and is underpinned by the assumption that when learners know what they are learning, how well they are learning, and how they can learn better, they are likely to become independent learners (Bell 2005; Bell and Cowie 1999; Willis and Cowie 2014). It is further elaborated by Klenowski (2009) as follows:

> Assessment for learning is part of everyday practice by students, teachers and peers that seeks, reflects upon and responds to information from dialogue, demonstration and observation in ways that enhance ongoing learning. (p. 264)

Formative assessment is influenced by classroom factors including student-teacher relationships, the physical set-up of the classroom, and learning opportunities provided by the teacher (Cowie 2000). It has been embraced internationally as a tool for enhancing learning and improving achievement (Klenowski 2009). As a globally popular intervention by teachers, formative assessment has been understood, applied in the classroom, and researched extensively. Teachers have been urged to provide students with feedback on their progress and suggestions for improving their learning (Brookhart et al. 2008; Heritage 2007; Nolen 2011; Wiliam 2006). Smith et al. (2014) argue that changing pre-service teachers' beliefs about the enabling power of formative assessment is essential to bring about change in classroom practice.

Inevitably, there are multiple understandings and varied applications of ideas, giving rise to many issues such as: tension between effective pedagogical approaches and testing for accountability (Black et al. 2011; McAdie and Dawson 2006); teacher unease about the pressure of external testing leading to adopting ways of teaching which are in conflict with their beliefs (Black et al. 2003; Moeed and Hall 2011); teachers using 'feedback to maximise reliability of summative assessment' (Shepard 2005, p. 7); how teachers practise assessment *of* learning, *for* learning and *as* learning; and continued teacher use of tests for formative practise for summative assessment (Hume and Coll 2009).

Other related issues include: misunderstood and misinterpreted practice of teachers' use of a mock *exam* or *trial run* just before the summative internal assessment, which is called *formative assessment* by some teachers (Hume and Coll 2009; Moeed 2010), and 'Frequent testing of levels achieved and then focusing on the deficiencies in order to meet the next level' (Mansell et al. 2009, p. 23). However, Black et al. (2011) have found that validity of teachers' summative assessment can be raised through teacher professional development.

Formative assessment was practised as a mock examination or trial run with the study class.

4.1.3 Summative Assessment

Summative assessment is assessment *of* learning (Crooks 2004). It is used to determine the effectiveness of teaching and learning. The purpose is to monitor educational progress or improvement. It usually comes at the end of the teaching and learning cycle and mostly provides a judgement about the learner either as a ranking within the group (normative), or as success against pre-set standards (standards based) (Bell 2005). In Crooks' (2004) view, although written feedback is not provided in this form of assessment (for example, in an external examination), it does influence future learning. Success in the examination communicates to the students (in the form of marks and grades) their areas of strength, which in turn leads to choices made for future learning.

Preparation for the summative assessment was the focus of learning in the study class.

4.1.4 Assessment for Evaluation

In the educational contexts, assessment for evaluation is used to ascertain effectiveness of a particular teaching and learning programme. The results are used for policy purposes and to establish that the resources put into teaching and learning are being used effectively. Assessment for evaluation also includes national monitoring programmes, in other words it is about accountability (Crooks and Flockton 1996). Scriven (2003) defines educational evaluation as the process of determining the merit, worth, or significance of things.

4.2 Assessment in Science for Year 11

In New Zealand, since the staged introduction of the NCEA in 2002, students have been assessed through a combination of internal and external assessment. Successful completion of 80 credits qualifies students with NCEA at level 1. Students then build on level 1 to achieve levels 2 and 3. Achieving level 3 is designed to acknowledge achievement across a range of learning areas and provides an advanced foundation for further study or employment (NZQA 2010; Phillips 2006). In year 11 science, a number of achievement standards are offered, some internally assessed and others externally assessed. NCEA allows schools the flexibility to design courses that address the needs of their students using a combination of internal and external assessment units and achievement standards (Hipkins et al. 2004). In the study school, one internally assessed achievement standard, carrying out a practical investigation (the focus of this research), and five externally assessed achievement standards were offered. The external achievement standards were in physics, chemistry, biology, geology and astronomy, and organic chemistry.

4.3 Results

Students' experience of formative and summative assessment in the study class and their views about the process of preparing and being assessed are described below.

4.3.1 Formative Assessment in the Study Class

The school had a practice of students carrying out a complete investigation to learn the process they would have to follow for their internal assessment of science investigation. Each year 11 class carried out this formative assessment which the teachers called a 'trial run' or 'mock exam'. This investigation is reported in Chap. 2 as 2.5.2. Students completed their plans and handed them in for the teacher to mark and provide feedback. Only 11 students handed in their plan for marking.

First, a group of students who appeared to be on task was observed:

> This group had set up three cups made of metal, glass, and paper with hot water in each and were taking readings at one minute intervals. They had the table set up in their workbooks and were recording their reading. Interestingly, they took the thermometer out from one after taking the reading and immediately putting it into the other and taking the reading. They were not aware that this would affect the accuracy of their readings. (Observation notes)

Another group of students sitting at the back was asked why they were not doing the investigation. Henry said that he knew which cup would keep the heat in "so what's the point?" Henry's group sat and talked and at the end of the lesson they had no data to record.

The focus group students said that they already knew the science concepts they were to learn through doing this investigation. The teacher's purpose for doing this investigation was formative, to give students an opportunity to practise an investigation and learn its terminology, and to become familiar with the process they would follow for the final assessment. The students did not pay attention, so the teacher, and students did not have a shared understanding of the purpose of doing this investigation. This became apparent when one student said that he did not want to do it because he had already done it in the previous year (see Sect. 2.5.2).

When students in the focus group were asked how this investigation helped them to prepare for the final assessment, the following discussion took place.

Bob: It showed us the lines which we had to follow to write out the method. How to use our time?

His response helped the others to remember doing this investigation:

Pip: And what we were looking for when we were doing it.
Researcher: Do you remember your teacher marked and returned your
                              plan? Did anyone use that feedback that the teacher gave you?
Bob: No.
Researcher: Why?
Jake answered instead: I tried to, but I didn't improve my mark, but I'm still
                              happy because I passed. (Focus group interview)

4.3.1.1 Student Questionnaire Results

Students were surveyed after the summative assessment of science investigation.
Twenty two students responded to a questionnaire about the formative assessment
they had done to prepare for the summative assessment, and what they had learnt
from it.

Preparation for internal assessment (Question 1)

Students were asked to identify up to three investigations they had carried
out and tick a box on a four-point rating scale to indicate their preparedness
(Table 4.1).

The students were then asked why they had chosen the response they gave.
Most students reported that rolling a marble down a ramp prepared them well or
very well for assessment (15 of 22). Not all students wrote a comment, but from
those who did, some students found it useful because it was the same as the one
they did for the assessment (3), find out how surface affects speed (6), and how
variables affect outcome (3). Five students found the body power investigation
helpful but did not elaborate, and two found it helpful for remembering the
formula (for working out power).

Students identified what they had learnt from the formative assessment for heat
retention by cups made of different materials. They said different materials retain
heat differently, and black things keep heat in (6); learnt or practised skills like

Table 4.1 Student preparedness for AS1.1

Investigation	Respondents (n)	Scale			
		Very well	Well	OK	Not very well
Rolling marble down a ramp	22	2	13	6	1
Comparing the power of different students running up a flight of stairs	13	0	3	6	4
Comparing how well cups made of different materials could keep heat in	14	0	3	4	7

measuring, graphing, and writing up experiments (3). There were a few negative responses: they did this last year in year 10 (4); it was boring (1); and they already knew what would happen (1).

Usefulness of formative assessment

Of the 20 students who answered this question, two found the formative assessment very useful, eight found it useful, and eight said it was somewhat useful. Two indicated that they did not find it useful at all. The two students who found formative assessment very useful thought it reminded them that their plan should be really clear so someone else could follow it and do the investigation. Another student who found formative assessment useful said, "Formative assessment told us all the things like variables and stuff we have to control, like having the same amount of water, the same temperature and things".

Helpfulness of teacher feedback for formative assessment

Question 3, was an open-ended question that specifically asked how teacher feedback was useful in preparation for the final assessment for AS1.1. Two of the student responses were coded for this question to which not all students responded. Most respondents indicated that the feedback was useful (19). Some found the feedback helpful in terms of writing the report (5); some found it helpful to have the template and knowing what they needed to put in each section (5); some found the feedback on planning useful (4); others said they gained confidence when the teacher told them that they had done well in the formative assessment (2).

The second part of the question asked about the usefulness of formative assessment for learning science. The students did not say how the feedback helped them with their science learning but said things specific to assessment that included: the feedback helped them to improve their report writing; the assessment improved their understanding of what they needed to learn; it helped them to focus on learning. Although only 11 students had handed in their formative task for marking and received feedback, the question allowed inclusion of other feedback useful in preparation for the formal AS1.1 assessment.

In sum, this investigation was designed to give students the opportunity to find out what they needed to do for their formal assessment. However, a third of the class did not put in the effort perhaps because some of them had already done this investigation in year 10. Only 11 students had handed in the planning task for this formative investigation and some reported that they did not use the teacher feedback to improve. In terms of science learning, it would appear that the students saw the investigation as identifying science knowledge (finding 'the answer'). They appeared to not appreciate the point of investigation as learning a process for finding out science knowledge and ideas.

4.3.2 Summative Assessment in the Study Class

In year 11, students carried out a science investigation which was assessed through Achievement Standard AS1.1 for credits towards the NCEA. This achievement standard is worth 4 credits and there are four possible grades: Not Achieved; Achieved; Achieved with Merit; and Achieved with Excellence. The task is internally assessed and students were allowed three, 1-hour sessions to do this investigation.

4.3.2.1 Illustrative Example, Assessed Investigation: Watch that Car Go!

'Watch that car go' was a moderated task available from the Ministry of Education. Students were given the following information in a hand-out for this assessed investigation.

Background Information:

Watching children play with their toy cars, students noticed that the cars seemed to travel at different speeds depending on the slope they were released on.

Task:

In this investigation you are to plan, collect, process and interpret information, and present a report to find out how the slope of a ramp affects the distance the car goes along the flat (table or floor). This investigation is a "**fair test**" investigation. (The emphasis is original).

Students had to plan the investigation during a normal class lesson on Friday before the start of the mock examination week. The teacher marked the planning task. Data gathering took place during the mock examination week in the college hall. Many secondary schools in New Zealand set aside a week in which the school exams are held; they are often called mock exams because they are a preparation for the real exams. The participating school held the internal assessments that contributed towards the final exams during this week.

Along both sides of the hall, 30 desks were set up so that students could do a written science mock examination for the externally assessed achievement standards. In the middle of the hall 30 two-metre long carpet strips were placed so that 30 students could carry out the data gathering. Other equipment placed next to each carpet strip included one-metre long pieces of wood to be used as ramps, three 10 cm blocks of wood to increase the height of the ramp, a toy car, a metre ruler, and a stop watch.

The teacher returned the marked plan to the students to use for carrying out the investigation. They divided the class into two groups and gave the following instruction:

> People, this is the last talking you do today. Group one, you will sit at the desks along the side and do the mock exam for the first hour. Group two go and sit next to one of the carpets set out in the middle. (Waits for the students to sit next to a station set out.) Group two you will carry out the practical investigation. The paper you are writing on will be collected and returned to you in class. You will be drawing graphs in the next lesson. After one hour you will swap over. (Study class teacher)

Six students were absent when the class had done the planning task. The teacher had organised for them to come and complete the planning task an hour before the assessment. They were put in group one and the teacher marked their plan while they did the mock examination. As their plans were workable, none of the students were given the plan that the teacher had to give to those who did not have a workable plan, and so they would have been disadvantaged.

According to an observation record, the researcher noted that:

> It is pretty easy to work out as the students sitting on the side doing the mock exam could look at what the students in the middle were doing. Two who were sitting on the side watched Bob and when it was their turn followed what he had done. This was different to what they had put down in their plan. (Observation notes)

Students processed the data and wrote the report in their science class a week after they had collected the data and after mock examinations were over. The assessment was returned to the class in the last lesson before the start of the school holidays. There was no opportunity for students to discuss their assessment.

4.3.2.2 Students Goals for AS1.1

Following the school policy, students had set goals for science achievement and achievement in AS1.1. Table 4.2 shows student grade goals for AS1.1 and the grades they achieved.

The students' results were far lower than the goals they had set for themselves. None gained "Excellence" (whereas 25 % had set that goal), 17 % gained "Merit" (33 % had set this goal), and 70 % gained "Achieved" (and only 42 % had set this minimal goal). A further 13 % did not achieve the standard, a disappointment for the students as none had expected a "Not Achieved" grade.

4.3.2.3 Student Focus Group Interview Results

The purpose of this interview was to find out specific assessment related information. The limitation of this interview was that it was done 14 weeks after the completion of AS1.1, which had implications for the reliability of the data. Students were reminded that normally in their class they work in groups to do practical

Table 4.2 Comparison of students' grade goals with grades achieved for AS1.1

Possible grades	Students' grade goals (%)	Students' grades achieved (%)
Excellence (E)	25	0
Merit (M)	33	17
Achieved (A)	42	70
Not achieved (NA)	0	13

work. They were asked how they found the assessment where they had to plan and carry out the investigation by themselves:

Bob: Even though I prefer doing it by myself it was kind of interesting going from being part of a group to doing it by myself because you had to see everything through by yourself, instead of having someone else's ideas.

Jake: It was different. It wasn't better or worse it was just different.

Ed: I think that people who are less dominant might have problems with that. But it was pretty (good) for the most part. (Focus group)

Students were asked how they found the process of carrying out the investigation in the hall with half the class doing the written examination while the other half carried out the investigation. This is how Bob described his experience:

> I think it was actually a good idea, because if you had everyone doing the exact same thing at the exact same time it could be a bit confusing. When I got up I was like one of the first to get down, but it was kind of free because you were standing in the middle of a hall and no-one else. Looking around and no-one else is moving. But it's a lot easier because you have all your stuff that you need laid out for you, and no-one else is trying to take it, so there's far less people to worry about and you can just get on and do it.

Jessica, who did the written mock examination, before the investigation said:

> I thought I'd be a bit distracted by everyone else getting up and doing their thing. But I wasn't, I just sort of tuned them out and got onto what I was doing.

Students were asked if they were worried about the assessment while they were doing the written examination. Simon said:

> It was kind of like the practical is worth credits. It's first priority. Do that, you know.

When asked if it was about credits, Jake said "Pretty much"; the others laughed and agreed. Ed added "I didn't even know the other was just mock". They were probed about what they had learnt from doing the assessed investigation. Pip, who said that she does not remember much science, commented she remembered this investigation: "I guess it stuck better because it wasn't theory, it was practical. It was easier to remember". When they were asked how they could have prepared themselves better for the assessment, Jake and Bob said:

Jake: Well I'm just glad I passed really. I don't think I could have done anything really better than I did, well me personally, Yeah.

Bob: Yeah the answers were simple really, you just had to revise them all and know what you were going to go into.

They were told that they did not have to share their results but asked if they were satisfied with what they had got? Two students said they did not remember what they got, one said he was happy. Jake said "Anything over an Achieved is good". Bob's response was unexpected:

> You cannot complain from passing, if you wanted an Excellence or whatever. If you're passing you're still getting the credits or whatever, so it doesn't really matter in the long run.

Jake said "It's not if you pass it's how you pass". This is interesting as in class Jake was easily distracted and often off task and Bob was focussed, always keen to answer questions, and having the right answers. His response was uncharacteristic. In the focus group he was the only one who described seeing a relationship and a pattern in the data during the assessment:

> It was great, every time I put another block to raise the height of the ramp 10 cm the car travelled 30 cm more. I did this many times and each time the same thing happened, how good is that?

He was disappointed but accepting of his grade, adding:

> And I got what the person marking the sheet thought I should have got, and I'll respect what they think. I think I maybe should have got something a little bit better. But if that's what they thought then that's what they thought, it's up to them, they're marking it.... I think I was a bit too distracted really. I didn't really do my best. Maybe if I was a bit more prepared, and a bit less worried and everything, I would have done quite a bit better.

In the focus group interview, when students were asked what they had learnt through doing this investigation, all their responses were about the science concepts rather than the process of investigation. Two responses illustrate this:

Bob: How gravitational potential energy turns into kinetic. How friction goes against it?

Jake: Depending on the different surface that you're rolling them on it depends on how far and how fast it (toy car) will travel

In this interview, the students were comfortable in talking openly and trusted that what they shared would not be shared with the teacher or other students. They were spontaneous in their responses. Their responses suggested that they were honest about their preparation or lack of it. Bob was clearly disappointed with his result for AS1.1 but did not blame the teacher or the process. Ed was his usual self; he knew he had not worked and was a bit lucky to get an Achieved. He shared with the group that he had failed all other standards but tried to put across that he did not care. Interestingly, though pass and fail are not in the NCEA grades, students were still talking about passing or failing if they achieved or did not achieve.

4.3.2.4 Student Questionnaire Results

This questionnaire was administered in term 3 (September) by the researcher after the AS1.1 assessment had been completed and students had received their grades. The students took 20–25 min to complete it. They were not asked to write their names on the questionnaire.

Student satisfaction with internal assessment

This question was to determine student satisfaction with the results. It was a 4-point rating scale on which students had to indicate their response. Only one student was completely satisfied, eight said they did as well as they could, six said they could have done better, five were not satisfied, and two did not respond.

Responses to what students themselves could have done to improve their grade for AS1.1 indicated that they were aware of how they could have achieved a better grade. The most common response was that they could have studied or revised (10); some said they could have written a better report (3); others thought making repeat trials (2) and drawing a graph would have been useful. Two students said they should have tried harder (2).

Enjoyment in doing the science investigation for AS1.1

Students were asked about their enjoyment of the investigation they did for formal assessment. Responses included: getting credits (2); passing (2); playing with the toy car during the actual assessment (2); it was not too hard (3); they had all the gear (2); everyone could do the work (1); and it was quieter (1). Negative responses included: I would rather do chemistry (1); I don't like physics (4); and I don't like science and don't enjoy any aspects of it (1). No comment was given by six students.

Aspects of the science investigation for AS1.1 not enjoyed by students

The purpose of this question was to find out aspect(s) of the investigation for formal assessment that students did not enjoy. Two students indicated that they enjoyed it all. Others said: having to do it (2); writing the discussion (4); because it was a test so you had to get it right; the fact that it involved math (1); they did not know everything (1), and they had to understand what was going on (2); experiments were boring (1); and everything in science is boring (1).

Understanding science ideas through the science investigating for AS1.1

In this question students were asked about how science investigation helped them understand science ideas learnt in class. Eighteen out of 22 students responded to this question. One student said they found it very helpful, eight found it mostly helpful, nine said it was somewhat helpful, five did not find it helpful, and four did not respond. Their responses showed that most students found this investigation helpful to some degree in their learning of science.

Eight of the 18 students provided comments. They indicated that it helped them to understand the formulae they needed and how to write an evaluation. Others provided general comments such as: helped me to learn; made it easy because when they got to "do" the investigation it helped them to remember; "because it lets me see the concept rather than having to 'see' it in my head" (Respondent 8); and because investigation "proves the ideas and actually shows them" (Respondent 13). All five of the students indicating investigation was unhelpful gave comments which included: "cause it was stupid, if a ramp is steep something will go faster" (Respondent 17); because the class mucks around (2); because they learnt more out of the book work than in the investigation (1); and "I already knew this from year 9 and 10" (Respondent 3).

4.3.2.5 Summary of Questionnaire Results

Most students remembered the investigations they had done. More remembered the investigation of rolling the marble down a ramp than other investigations. Rolling the marble down the ramp was very similar to the one used for assessment of AS1.1 where the students had to roll a toy car down a ramp. The difference between the marble and toy car investigation was that in the former they investigated the effect of surface on the distance travelled by the marble whereas for the latter they investigated the effect of increasing the height of the ramp.

Most found the formative assessment and the feedback provided by the teacher useful or somewhat useful. The reasons why they found the feedback useful were mostly about the reporting or skills aspects of investigation. Fewer than half the students were satisfied with their results, some attributing this to lack of effort on their part.

4.4 Study Class National Certificate of Educational Achievement Results

The students and school had given consent and access to the NCEA results for these students for achievement standards assessed internally and externally. The school offered physics, chemistry, biology, and astronomy achievement standards. In the year of the study they also offered the organic chemistry achievement standard to students taking science. Practical investigation in the school was assessed through science AS1.1. Students were able to choose which external achievement standards they wanted to enter.

The results for internal and external assessment showed that students performed better in the internal assessment of science investigation with four students achieving a Merit grade, 17 an Achieved grade, and only three did not achieve. None of the students received an Excellence grade. Only Jessica and Phil achieved in all standards and Harry and Linda did not achieve in any of the standards they entered. Seven students—Jeff, Craig, Len, Ken, Ed, Mili, and Henry—only Achieved in internal assessment. Overall, over half the students attained fewer than 12 of the 24 credits offered (Table 4.3).

In the study year, the national and study school average grade for Achieved or better for AS1.1 was 83 % and the study class average for AS1.1 was slightly higher at 88 %.

Table 4.3 Study class internal and external assessment results for NCEA Level 1

Achievement standards	Science					Chemistry	Total
	AS1.1	AS1.3	AS1.4	AS1.6	AS1.7	AS1.7	
	Science investigation	Biology	Chemistry	Physics	Astronomy	Organic chemistry	
No. of credits	4	5	5	5	2	3	24
Internal/ External	Internal	External	External	External	External	External	
Jessica	M	A	M	M	M	A	24
Andy	A	A	N	A	N	N	14
Bob	A	N	A	A	A	A	19
Jeff	A	N	N	N	N	N	04
Craig	A	N	N	N	N	N	04
Amy	A	A	A	A	A	N	21
John	A	N	N	A	N	N	09
Len	A	N	N	N	N	N	04
Ken	A	Y	Y	Y	Y	Y	04
Jake	M	N	A	N	A	N	11
Emily	M	A	A	A	A	N	21
Phil	M	A	A	A	A	A	24
Nikki	A	N	N	A	N	N	09
Simon	A	A	N	A	N	N	14
Dan	A	N	A	A	A	N	21
Susan	A	A	A	N	A	N	16
Ed	A	N	V	V	N	V	04
Mili	A	N	N	N	N	N	04
Robin	N	N	V	A	N	N	05
Linda	N	N	V	N	N	N	00
Jamie	A	N	A	N	N	N	09
Pip	A	A	A	N	N	N	14
Harry	N	N	N	N	N	Y	00
Henry	A	N	N	N	N	N	04

Key: *A* Achieved, *M* Merit, *N* Not Achieved, *V* not appeared, *Y* not entered

4.5 Validity, Reliability, and Manageability

The guiding principles that underpin assessment are validity, reliability, and manageability. Validity is about 'fitness for purpose' and is an essential criterion for the 'worth' of an assessment (Hall 2007). Harlen (2005) contends that validity is about what is assessed and how well this corresponds with what it is intended to assess. Hall puts forth the notion of three types of validity that include face validity, content validity, and consequential validity. He says that validity depends on a number of factors:

- the extent to which the purposes of an assessment are clearly described and the tasks used to judge students' progress and achievement seemingly relate to these purposes (called *face validity*)
- the extent to which the assessment framework for a course/module samples appropriately the content and learning outcomes (called *content validity*)
- the quality of the assessment tasks in terms of their relevance, diversity, construction and clarity (an aspect of *content validity*)
- the extent to which assessment criteria have been clearly communicated to students (an aspect of *content validity*)
- the extent to which assessment tasks do not have harmful side-effects, such as creating unnecessary stress in students or promoting surface learning instead of deep learning (called *consequential validity*)
- the extent to which the marking procedures and feedback processes help students improve their future performance (an aspect of *consequential validity*). (Hall 2007, p. 5)

In Hall's view, validity issues can be addressed through pre-moderation by subject specialists. The difference between construct validity (Harlen) and content validity (Hall) is that construct validity is about the task used and whether it assesses what it is designed to assess, whereas content validity takes into account whether the task measures a wide range of knowledge and skills taught within the course rather than an aspect of the course.

For the assessment in the present research, assessment validity in the case of a practical investigation is an assessment of performance either partly or in its entirety (Roberts and Gott 2006). Roberts and Gott argue that the observation required for assessment of practical performance is not possible in mass education, which clearly was an issue for assessment in this study.

Manageability deals with practical considerations including affordability, access to resources, time, and workload for teachers. In Hall's (2007) view, standards-based assessment is problematic due to teacher and student time required in dealing with management issues that in turn affect validity and reliability. The school has to find effective ways of managing assessment. In this case the students did the planning in class in 1 week, did the data gathering during the examination week 10 days later, and wrote their report a week after that in class. The data collection was done in two large groups in full few of what other students were doing and raises the issue of reliability.

4.6 Summary

Some students believed that they had been adequately prepared for the assessment through the formative task they carried out in class. Fewer than a third handed in their plan and received feedback on what they needed to work on. Of those who did get the feedback, some were unsure if they had used it. It can be concluded that the formative assessment, which has the potential to improve learning, could have been more effective. The results of the internal assessment show that most students achieved in this assessment and at best it can be said that these students had learnt how to carry out a fair testing type of investigation. Generally, their ability to plan and carry out a science investigation is not known.

There is no denying that assessment has an important place in teaching and learning in school science. What is debatable is teaching *for* assessment and teaching *to* assessment as Cleaves and Toplis' (2007) research suggests.

References

Bell, B. (2005). *Learning in science: The Waikato research*. New York: RoutledgeFalmer.

Bell, B., & Cowie, B. (1999). Researching teachers doing formative assessment. In J. Loughran (Ed.), *Researching teaching*. London: Falmer Press.

Biggs, J. (2003). *Teaching for quality learning at University* (2nd ed.). Maidenhead: SRHE and Open University Press.

Birk, J. P., & Kurtz, M. J. (1999). Effect of experience on retention and elimination of misconceptions about molecular structure and bonding. *Journal of Chemical Education, 76*, 124–128.

Black, P., Harrison, C., Hodgen, J., Marshall, B., & Serret, N. (2011). Can teachers' summative assessments produce dependable results and also enhance classroom learning? *Assessment in Education: Principles, Policy and Practice, 18*(4), 451–469. doi.org/10.1080/0969594X.2011.557020.

Black, P., Harrison, C., Lee, C., Marshall, B., & Wiliam, D. (2003). *Assessment for learning: Putting it into practice*. Maidenhead: Open University Press.

Broadfoot, P., & Black, P. (2004). Redefining assessment? The first ten years of assessment in education. *Assessment in Education, 11*(1), 7–27. doi:10.1080/0969594042000208976.

Brookhart, S., Moss, C., & Long, B. (2008). Formative assessment that empowers. *Educational Leadership, 66*, 52–57.

Cleaves, A., & Toplis, R. (2007). Assessment of practical and enquiry skills: Lessons to be learnt from pupils' views. *School Science Review,88*(325), 91–96.

Cowie, B. (2000). *Formative assessment in science classrooms*. Unpublished doctoral dissertation, University of Waikato, Hamilton, New Zealand.

Crooks, T. (2004, March). *Tension between assessment for learning and assessment for qualifications*. Paper Presented at the Third Conference of the Association of Commonwealth Examinations and Accreditation Bodies (ACEAB), Nadi, Fiji.

Crooks, T. J., & Flockton, L. C. (1996). *Science assessment results 1995: National education monitoring report 1*. Dunedin: EARU.

Driver, R., Guesne, E., & Tiberghien, A. (1985). *Children's ideas in science*. Philadelphia: Open University Press.

Hackling, M. W. (2005). Assessment in science. In G. Venville & V. Dawson (Eds.), *The art of science teaching*. Singapore: Allen & Unwin.

Hall, C. (2006). *Planning the assessment for a programme and its components: A guide for tertiary level educators*. Wellington: School of Education Studies, Victoria University of Wellington.

Hall, C. (2007). *Qualitative research designs*. Wellington: School of Education Studies, Victoria University of Wellington (unpublished lecture notes).

Harlen, W. (2005). Trusting teachers' judgement: Research evidence of the reliability and validity of teachers' assessment used for summative purposes. *Research Papers in Education, 20*(3), 245–270.

Heritage, M. (2007). Formative assessment: What do teachers need to know and do? *Phe Delta Kappan, 89*, 140–145.

Hipkins, R., Vaughan, K., Beals, F., & Ferral, H. (2004). *Learning curves: Meeting students' needs in a time of evolving qualifications regime: Shared pathways and multiple tracks: A second report*. Wellington: New Zealand Council for Educational Research.

Hume, A., & Coll, R. (2009). Assessment of learning, for learning, and as learning: New Zealand case studies. *Assessment in Education: Principles, Policy and Practice,16*(3), 269–290.

Jones, E. J. (2007). *A portfolio for assessment of the practice of special education resource teachers*. Unpublished doctoral dissertation, Victoria University of Wellington, New Zealand.

Klenowski, V. (2009). Assessment for learning revisited: An Asia-Pacific perspective. *Assessment in Education: Principles, Policy and Practice, 16*(3), 263–268. doi:10.1080/09695940903319646.

Mansell, W., James, M., & Assessment Reform Group. (2009). *Assessment in schools: Fit for purpose? A commentary by the teaching and learning research programme*. London: Economic and Social Research Council.

McAdie, P., & Dawson, R. (2006). Standardized testing, classroom assessment, teachers, and teacher unions. *Orbit, 36*(2), 30–33.

Moeed, A. (2010). *Science investigation in New Zealand secondary schools: Exploring the links between learning, motivation and internal assessment in year 11*. Unpublished doctoral dissertation, Victoria University of Wellington, New Zealand.

Moeed, A., & Hall, C. (2011). Teaching, learning and assessment of science investigation in year 11: Teachers' response to NCEA. *New Zealand Science Review, 68*(3), 95–102.

New Zealand Qualifications Authority. (2010). *NCEA: How does it work?* Retrieved March 2, 2010, from http://www.nzqa.govt.nz/ncea/about/overview/ncea-thedetails.html

Nolen, S. B. (2011). The role of educational systems in the link between formative assessment and motivation. *Theory Into Practice,5*, 319–326.

Osborne, R. (1982). *Science in primary schools: Working paper 101. Learning in science project (primary)*. Hamilton: University of Waikato.

Phillips, D. (2006). The contribution of research to the review of National Qualifications Policy: The case of the national certificate of educational achievement (NCEA). *New Zealand Annual Review of Education,16*, 173–191.

Roberts, R., & Gott, R. (2006). Assessment of performance in practical science and pupil attributes. *Assessment in Education: Principle, Policy & Practice, 13*(1), 45–67.

Scriven, M. (2003). Evaluation in the new millennium: The transdisciplinary vision. In S. I. Donaldson & M. Scriven (Eds.), *Evaluating social programs and problems: Visions for the new millennium: The Claremont symposium on applied social psychology* (pp. 19–41). New Jersey: Lawrence Erlbaum.

Shepard, L. A. (2005). Linking formative assessment to scaffolding. *Educational Leadership, 63*(3), 66–70.

Smith, L. F., Hill, M. F., Cowie, B., & Gilmore, A. (2014). Preparing teachers to use the enabling power of assessment. *Designing assessment for quality learning* (pp. 303–323). Netherlands: Springer.

Wiliam, D. (2006). Formative assessment: Getting the focus right. *Educational Assessment,11*, 283–289. doi:10.1207/s15326977ea1103&4_7.

Willis, J., & Cowie, B. (2014). Assessment as a generative dance. In *Designing assessment for quality learning* (pp. 23–37). The Netherlands: Springer.

Chapter 5
Discussion and Conclusion

The Connectedness Between Motivation, Learning and Assessment

5.1 Emergent Themes

Seven themes emerged from the data that demonstrate the connectedness and complexity of learning, motivation to learn, and assessment within the context of science investigation:

Fair testing: Investigation in practice
Learning that science investigation process is linear and sequential
Learning through science investigation
Motivational influences of science investigation
Assessment as the driver of learning
Assessment reliability and validity issues
Assessment promoted low level learning outcomes

5.1.1 Fair Testing: Investigation in Practice

Students experienced mostly fair testing types of investigations. Fair testing was specified in *Science in the New Zealand Curriculum* (Ministry of Education 1993) as being achievement objectives in the Developing Investigative Skills and Attitudes integrating strand. The controlling of variables is an essential element of a fair testing type of investigation. Students did not have the opportunity to experience an open-ended investigation where they could seek an answer to a question. Millar (2004) argued that it is difficult for teachers to manage open-ended investigation where a large number of students are involved. There is a similar "over-heavy" emphasis on fair testing in the United Kingdom national curricula (Watson et al. 1999). However, Wellington (1998) reported changes in

© The Author(s) 2015
A. Moeed, *Science Investigation*, SpringerBriefs in Education,
DOI 10.1007/978-981-287-384-2_5

the United Kingdom so that there was less emphasis on controlling variables that led to less fair testing even though there continued to be one "template" model for science investigation. The template approach is what the students in the study class experienced. A downside to this kind of investigation is that students learn that science investigation is about following a series of steps to arrive at a correct answer rather than learning that investigations do not always provide a predetermined result.

A particularly influential factor for the emphasis on fair testing is that the assessed investigation for NCEA level 1 is a fair testing type of investigation. Fair testing is therefore required to be taught and, not surprisingly, is found to be the focus of teaching and assessment experienced by the students in this research. Although other types of investigation, such as pattern seeking, classifying, and exploration were included in the curriculum, they have not been specifically assessed for NCEA. The issue is that if other types of investigation are not formally assessed, they are less likely to be taught. More importantly, if students mostly experience fair testing they are likely to have a limited view of science investigation. The following theme highlights how students' learning was constrained by their learning experiences that were selected to prepare them for assessment.

5.1.2 Learning that Science Investigation Process Is Linear and Sequential

School science investigations are similar to how scientists investigate, but are not exactly the same. Scientists have a wealth of up-to-date knowledge and conceptual and procedural understanding when they embark on an investigation to answer a question of interest to them. Scientists investigate in many different ways depending on their particular discipline. They begin with the ideas they already have and the processes that they think might work. As the investigation advances, things may not always proceed smoothly, and they may have to make changes to the procedure, rethink their plan, and sometimes even abandon the entire approach and start again. It can be a messy business! In contrast, students seldom choose the question they want to investigate; often they do not have the knowledge, understanding and experience to bring to the investigation. It is the job of science education to help students build their conceptual and procedural knowledge during their schooling. Through experience, students can come to understand that science investigations are not about following steps to arrive at the "correct" answer. The students in this research experienced the kinds of investigations that led to the belief that science investigation is a linear process where one needs to follow steps, collect data, and write a conclusion. If science investigation appears to be a linear and sequential process then students are gaining an unrealistic understanding of the very nature of science investigation.

Hodson (2014), who has researched teaching and learning from practical work extensively, proposes four phases of a science investigation (inquiry in terms of school science investigation):

(1)   A design and planning phase, during which specific research questions are asked and goals clarified, hypotheses formulated (if appropriate), investigative procedures devised and data collection techniques selected.

(2)   A performance phase, during which the various operations are carried out and data are collected.

(3)   A reflection phase, during which findings are considered and interpreted in relation to various theoretical perspectives, conclusions drawn and justifications for those conclusions formulated and refined.

(4)   A recording and reporting phase, during which the procedure, its rationale and the various findings, interpretations and conclusions are recorded for personal use and expressed in the style approved by the community for communication to, and critical scrutiny by, others. (p. 9)

In all the observations made in the study class, the reflection phase was almost always missing and the students did not have the opportunity to reflect on their investigation. To understand the nature of science investigation, this is the critical phase when evidence is identified, critiqued and evaluated. The students collected the data and the lesson finished with the class leaving with the instruction to complete the write-up for home work. There was little evidence to follow up if they had thoughtfully considered their results, critiqued their design, and drawn evidence-based conclusions.

5.1.3 Learning Through Science Investigation

Students said they experienced "repetition", "practising fair testing", and "doing investigations that we had done before". Observation data showed that they repeatedly learnt about controlling variables, and gathering and recording data. It appears that they were trained to 'go through the hoops.' This training was reinforced by constantly using the template designed for AS1.1. Cleaves and Toplis (2007) found similar trend in the United Kingdom and that the students were aware of this practice. Toplis (2004) also reported that students in his study said that their teachers told them to find anomalous results and explain their results to get a better mark. This was also found by Keiler and Woolnough (2002) and Wellington (2005) in the United Kingdom.

Study class observations and focus group interviews suggested that students had rote learnt science facts, such as plastic is an insulator, and metal is a conductor of heat, but did not understand why hot water in a metal cup cools down faster than in a plastic cup. Similarly, they had the skills to set up an investigation to find out which cup kept the water hot for longer. Students had procedural knowledge to measure temperature and they had learnt the need to measure the temperature accurately at set intervals. However, they did not demonstrate procedural understanding that if they took the thermometer out and put it in again, they needed to give it time before taking the reading. Their learning approach focused on perceived skills needed to perform the investigation rather than understanding the application of the skills learnt.

It seems that procedural knowledge rather than procedural understanding, along with conceptual learning, were deemed appropriate preparation for assessment. In Abrahams and Millar's (2008) view and according to Roberts (2009), both conceptual and procedural understandings are needed to carry out science investigation. Instead of developing these two kinds of understandings of investigation, students in this study were trained to perform in the assessment of science investigation. Millar's (2004) theory that students need to form links between the domain of objects and the domain of ideas essential for understanding were not taught or scaffolded in the study class. The *Science in the New Zealand Curriculum* (MoE 1993) requirement of evaluating the procedure and findings, which encourages a deeper approach to learning, was not evident.

Students, when asked about the science they had learnt, almost always talked about the science concepts; for example, when asked what they had learnt about energy they talked about gravitational potential energy changing into kinetic energy. At times, when they did bring up learning from science investigation, they were more focussed on the process they had to follow for assessment of investigation rather than the purpose of doing the science investigation. There was little mention that would indicate the development of procedural knowledge or understanding of the very nature of science investigation. On several occasions students said they remembered science ideas when they could see it happen before their eyes. Indeed some were able to relate what they saw to real life; for example, when Harry burnt diesel and saw a piece of carbon float through the air, he could "see" how diesel was a pollutant. Lederman et al. (2014) state that "students will best learn scientific concepts by doing science" (p. 291), a claim that is challenged by Hodson (2014). Hodson argues that this is promoting the outdated "discovery approach" where students may or may not "discover" anything. Hodson asserts that learning science is too important to be left to chance. "When left to their own devices, students may fail to reach the kind of conceptual understanding teachers seek, but when students are too closely guided or directed, the activity ceases to be 'doing science'" (p. 2). He promotes a much more explicit approach both to learning science content and to learning NOS content.

5.1.4 Motivational Influences of Science Investigation

Rotgans and Schmidt (2014) contend that there is a close connection between interest and learning. It follows that most believe that the greater the interest in a topic, the more willing the learner is to learn.

Task engagement was high during the data gathering phase of most investigations and during other practical tasks in the science class. This may be explained by situational interest which, according to Palmer (2009), is short-term interest generated by a specific situation such as a spectacular demonstration, a practical activity that has a 'wow' factor, or one that has novelty. Illustrative examples of situational interest from the study class included students doing a number of exploratory investigations with dry ice and energy changes. In the case of the study school, situational interest could perhaps explain some students' interest in the setting up of the assessment of AS1.1 in the school hall as exemplified by the focus group students who said that the gear was set out, it was in working order, and "no one was taking away your gear". Although the teacher and the researcher can interpret engagement or enjoyment as motivational, it is only the students who can say if practical work is motivational and for what reason. Students' reasoning for motivation was identified through focus group interviews, and from informal discussion with the researcher, as well as student surveys. Enjoyment was reported by students to arise from interest, novelty, variety, making a personal input in the way practical work was conducted, being able to work with their friends, and being able to move around. The enjoyment of working with their friends was also a motivational reason offered by students in the focus group. The preference to "work with their friends", "help each other" and "work in groups" was evident in the statements students selected in the preferred version of the SLEI (Fraser et al. 1995, see Chap. 2). Palmer (2009) found that students enjoyed activities that allowed them to work with their friends and to move about. On a negative note, investigation was described as providing an escape from writing, which included copying from the board and completing worksheets.

Palmer (2009) posits that students are interested in tasks that may have personal relevance. The results of the SLEI from the study class showed that students had a preference for pursuing something of personal interest, but this did not often happen as all investigations and practical work were either teacher directed or were tasks from workbooks. The students seldom investigated something of interest to them. There were, however, glimpses of choice-dependent interest as demonstrated by one girl who was allowed to go outside for the "Rolling a marble down a slope" investigation task, on her request, and investigate if the marble rolled further on astro-turf. This was a memorable event for her, which she shared in the focus group interview.

Palmer (2009) and Alexander et al. (1994) have reported novelty as a motivating factor for doing investigation that arouses situational interest. The influence of novelty was observed in the study class where there was a 'wow' element to the task that showed student enjoyment, for example, the excitement when students burnt diesel fuel and could see bits of carbon float through the air. In contrast, repetition can be demotivating as evident in the teaching scheme of the study school where there was an overlap of up to 90 % between the practical activities in some topics in year 10 and year 11. Due to this repetition, novelty was not often experienced by the students in the study class. Some (5–8) students in the study class did not participate in some of the investigations claiming that they were boring because they had done these before and they were not of interest to them, which is consistent with Rennie et al.'s (2001) view that adolescents find science as a dull and boring subject that fails to motivate them.

Students in the study class said that they wanted to learn when they found the activity interesting. Students in the focus group reported being able to remember concepts learnt during investigation when they had enjoyed the experience and seen the results "happen before their eyes" such as a bimetallic strip bending as it was heated. Similar findings in relation to memorable events have been reported by Alexander et al. (1994) and Eccles and Wigfield (2002).

Scholars argue that motivation is a force that "induces learners to persist with a task" (Izard 1977) and is a pre-requisite and co-requisite for learning (Palmer 2009). Students in this study were motivated to do practical work when it had novelty; they enjoyed working with their friends; and they found practical investigation an attractive alternative to copying notes. There was agreement among students that NCEA credits and grades were a strong extrinsic motivating factor.

Interest and enjoyment are separate feelings but in fun, both should occur together. Interest is the exploration of what is novel and intriguing, being fascinated and getting caught up, whereas enjoyment is the satisfaction that comes from participating in the activity. Joy is the confidence that comes from knowing that one can cope with the problems successfully.

5.1.5 Assessment as the Driver of Learning

The internal assessment of science investigation was central to all practical work and science investigations until the assessment took place in the middle of the year. After the assessment, there were no further investigations carried out in the study class for the rest of the year. It appeared that the first half year was preparing for the internal assessment, and the second half year was devoted to preparing for the final examinations when the other achievement standards were assessed. In the study class, there was little evidence of diagnostic assessment, and formative assessment was limited to one "complete" investigation as a "trial run".

Most investigations that students did were of the fair testing type and, as said earlier, it was a matter of using a template to plan and carry out an investigation following set criteria. The results showed a very high level of achievement in the internal assessment.

5.1.6 Assessment Reliability and Validity Issues

Assessment reliability is about consistency or accuracy of results across assessors and over time (Hall 2007; Harlen 2005). For assessment of investigation, high "reliability would entail students getting the same results all the time irrespective of when the assessment is carried out and who marks it" (Harlen, p. 246). Being trained to achieve in assessment may enable students to rewrite the same answers and get the same result if the same assessment task is used under the same conditions. This would not be an appropriate indicator of student understanding of science investigation and may compromise the validity of assessment. Some teachers in the study school said that even though a student may get an Achieved grade for AS1.1, they could not say if that student was capable of achieving it in another assessment. Hume and Coll (2008), in their case study, found students who had been able to carry out a fair test involving rates of chemical reactions were unable to carry out an investigation in a physics context of simple pendulums.

Reliability can be increased by doing five to ten assessments of investigation in different contexts and taking an average (Gott and Duggan 2002); however, this is not a realistic option in New Zealand as it would be too time and labour intensive and could mean the curriculum would not be covered. Reliability can be increased by using a template and tightening the criteria (Gott and Duggan 2002; Hodson 1993); both are features of investigation used for AS1.1. Very easy or very difficult assessment tasks are likely to be less reliable (Kraska 2008). The high level of student achievement in AS1.1 in the study school and nationally (both 83 %) suggests that the assessment task was comparatively easy for students in year 11, and so possibly the task was not sufficiently reliable. Another explanation could be that it was poorly implemented because students were trained and given plenty of direction.

Although the criteria have been tightened, the implementation does not appear to reflect this change for the most commonly used tasks for AS1.1. Both the task and marking schedule were available on the New Zealand Qualifications Authority website (NZQA 2005), and are easily accessible to students. According to the study class teacher and New Zealand Qualifications Authority statistics, the same task was used nationally for over eight years. Potentially, students could find the task and prepare for it and write the expected answers indicated in the marking schedule to get an Achieved, Merit, or Excellence grade. There was no evidence that this was actually taking place in the study class. Further, the study class teacher suggested that students would talk to other students and find out what they needed to write to get an Excellence grade. In the study school where assessment took place in three lessons spread over two weeks, students had ample time to find out specific information required and use it in their report before marking and feedback occurred.

Validity is an essential criterion for the worth of an assessment. Assessment validity is about how well the task(s) assesses what it is intended to assess (Harlen 2005). A valid assessment provides information which is useful, appropriate, and accurate. There are several types of validity of which face, construct, content, predictive, and consequential validity are relevant. For the assessment task "Watch that car go", the judgement statement provided for the marker relates to the purpose of the assessment. Therefore, in respect of purpose, the task "Watch that car go" demonstrates face validity. Similar face validity is shown for other moderated tasks. Construct validity is measured by aligning the knowledge and skills that the task ought to measure to what it actually measures. In the case of assessment for AS1.1, for example, the task "Watch that car go" is designed to assess students' ability to carry out a fair testing type of investigation with direction and allow for differentiation between Achieved, Merit, and Excellence. The task measures planning, controlling of variables, gathering, processing, and interpreting data and writing a report. Therefore, the assessment task is strong on construct validity for a fair testing type of investigation as defined by the NCEA. However, the task is less valid for a fair testing investigation that includes focussing and problem-solving, as defined in *Science in the New Zealand Curriculum* (MoE 1993) or for science investigation more holistically.

The "Watch that car go" assessment task was also likely to have limited predictive validity because the results of the assessment, in the view of the teachers in the study school, would not ensure that a successful student could carry out a science investigation in a different context or carry out a different kind of investigation. For example, the strong focus on training students is likely to de-emphasise some of the higher intellectual processes that promote transfer of learning across contexts. Fair testing is only one of the many types of investigations (Watson et al. 1999). The implication is that this focus in AS1.1 leads to teaching practices that may not appropriately reflect the learning outcomes required by the curriculum in the domain of 'investigation'. These learning outcomes at level 6 include focussing, planning and carrying out a problem-based or topic-based open-ended investigation, and developing an understanding that investigations require an iterative process. Neither focussing nor iteration are identified in the fair testing investigation assessed through AS1.1 and therefore this assessment of investigation had low content validity at level 6 of the curriculum. Additionally, since fair testing tasks favour chemistry and physics and any one task can only address one science discipline (and sub-disciplines within that), AS1.1 can be said to have low content validity within the domains of science.

The assessment of science investigation, as required by NCEA and implemented in the study school, has had negative side-effects, including: encouraging a surface approach to learning; providing a narrow focus on fair testing types of investigation; students being given the training to perform such an investigation; and teachers having limited use of formative assessment and feedback. These negative side-effects highlight issues of consequential validity. The assessment of investigation, as prescribed and implemented, may be doing harm and is therefore open to challenge in terms of consequential validity (Crooks 1993 cited in Hall 2007).

However, on a more positive note, assessment reliability is raised through the standardisation of requirements, practices and conditions for the administration of assessment, such as the structured template required by the New Zealand Qualifications Authority. The increased reliability does go some way to building the consistency and predictability that school-based assessment of student learning should aim to achieve.

There were a number of issues that influenced the validity and reliability of AS1.1 in the study school. Evidence from observations and teacher interviews in the study school showed that formal assessment of the hands-on practical phase of investigation was carried out for a large number of students at a time and in a public space where each student worked in full view of other students. Some students changed their plans from those marked and returned to them by their teacher because they could see what the others were doing during the assessment.

Based on the preceding commentary on the validity and reliability of the fair test investigation, it can be inferred that when constructed carefully, administered appropriately, and interpreted properly, assessment of science investigation could provide an in-depth window into how students apply their knowledge and skills to carry out an investigation (Harlen 2005). The NCEA requirement of assessment of a single fair testing type of investigation using a tightly structured task is likely to have increased assessment reliability but places constraints on validity.

5.1.7 Assessment Promoted Low Level Learning Outcomes

The NCEA assessment of science investigation resulted in teaching and learning of one type of investigation. The constraints of this type of investigation and the limitations of the standards for each level of achievement encourage low level learning outcomes that may result in lower order thinking. To explain what is meant by low level learning outcomes the taxonomy of the cognitive domain first put forward by Bloom et al. (1956), and more recently updated by Anderson and Krathwohl (2001), is used. This framework has six categories from the simplest to the most complex—knowledge, comprehension, application, analysis, synthesis, and evaluation. If the assessment requirements were aligned with Bloom's taxonomy, students who demonstrated that they had met the criteria for the standard, Achieved, appeared to be operating at the "knowledge" level. The judgment statements for the marking of AS1.1 for Achieved level required students to describe, identify, measure, record and select; unless these skills involved deeper processes than normally associated with their performance, they aligned with Bloom's knowledge level.

To attain a Merit, the standard requires "(p)rocessing of data to enable a trend or pattern (or absence) to be determined" and "a valid conclusion based on interpretation of the processed data that links to the purpose of the investigation". This required analysis and would be congruent with higher order thinking. However, in the "Watch that car go" assessment task, the acceptable response for Merit required accurate measurement of distance, more than two trials, and providing a valid conclusion based on the purpose of investigation such as, "(t)he greater the height of the ramp, and therefore the steeper the slope the greater the distance travelled by the car" (Appendix 2, assessment schedule). The Merit grade requires higher order thinking but the actual marking schedule stipulates a simple answer that may not require a lot of thinking and best fits with Bloom's comprehension level.

Further, the Excellence level answer requires discussion based on evaluation and justification that are higher order thinking objectives. However, as for Merit, the assessment criteria required repeatable results and enough readings to allow a "valid" trend. The science idea used as an example said "The higher the ramp the more gravitational potential energy the car had and the more that was converted to kinetic energy, therefore the further the car went before it came to a halt" (Appendix 2; NZQA 2005). If the students were working from first principles it would require proportional thinking, but the problem is that students may learn it by rote and recall it. For students who had been "trained" to perform using tasks almost identical to the assessment task, this assessment was unlikely to require higher order thinking even though the original learning may have done.

Classroom observations, including formative assessment, showed that in the fair testing investigation students did in class, some memorised and wrote answers that could get them an Excellence grade. In the sample assessment schedule provided for "Watch that car go" (MoE 1993), the criteria for Excellence in reporting said, "the final method chosen gave results that were repeatable … this also allowed us to see aberrant data and this data could be removed from our calculations." Acceptable examples of such answers in the study school were "I repeated the trials and took an average to get more reliable results," (Harry) or explaining an anomalous result as a measurement error on the student's part (MoE 1993). However, none of the students in the study class achieved an Excellence grade but several did get a Merit. Students wanted to know what they needed to write to get a particular grade; they were not asking for help to understand why they should be writing that answer. This focus on recall reduced the Excellence level answer to a much lower level of knowledge for most students, with the assessment task not requiring students to exhibit the higher order critical thinking skills. In New Zealand, Hume and Coll (2008) also found that "students were acquiring a narrow view of scientific inquiry where the thinking was characteristically rote and low-level" (p. 1201).

5.1.8 Assessment Encouraged a Surface Approach to Learning

Students in this study had a surface approach to learning investigation. Entwistle (2005) defines a learning approach as "what the students intended to learn and how they learn it" (p. 2). In a surface approach to learning the intention is to memorise facts and processes. Ideas are accepted passively; the focus is on the requirements of assessment, reflection on learning is minimal, and learners fail to recognise patterns and organise ideas. A surface approach to learning is often due to anxiety and fear of failure (Entwistle and Ramsden 1983). These were common concerns in the study class where students felt driven by the need to achieve credits for NCEA level 1. Very few students in the study class showed elements of a deep learning approach "which is linked to academic interest in the subject" (Entwistle 2005, p. 3). The study class students, when asked to set goals for their science learning and for AS1.1, identified performance goals, for example, achieving a Merit or Excellence grade rather than mastery goals such as learning something of personal relevance or interest. Such goals are indicative of a surface approach to learning in contrast to learning for personal relevance or interest which are mastery goals linked with a deeper approach (Eccles and Wigfield 2002). Conversely, some of these students may have been strategic learners who focused on what they perceived as important for assessment.

When undertaking a fair testing investigation almost all students in the study class had drawn tables in their books and entered data showing that they had taken several readings. When asked why they needed to take several readings, a number including Harry responded "to take an average", which they explained made their data reliable, but they could not explain what they understood by reliable data. Similarly, during another investigation a student set up a test tube to measure the energy released when different fuels burnt. The test tube was placed in a retort stand and had a thermometer in it. This student then proceeded to put the fuel he had collected from the teacher in a bottle top and put it under the test tube to light the fuel. When asked what he was doing he could not explain why he was about to heat an empty test tube. In this case the student was trying to reproduce the procedure followed by other students around him and missed the detail that these students had water in their test tubes. His approach to learning was minimal as he just wanted to get the investigation completed and did not appear to think about learning from this experience.

During the same investigation another student found that all the fuels she had burnt boiled the 20 ml of water she had put in the test tube. Her conclusion was that all fuels released the same amount of energy. Upon further probing she said it could be because her test tube was not clean but she was unable to see that if she had used a larger quantity of water (100 ml) she would have been able to measure the change in water temperature for each fuel burnt. Her response demonstrated a lack of reflection on the results which is characteristic of a surface approach to learning. The student's response to the test tube being dirty was in agreement with Cleaves and Toplis's (2007) findings in the United Kingdom that students had a rote response to anomalous data if they did not get the expected results.

The student workbook set out tasks in the template format required for AS1.1, like a recipe. The tasks were explained in detail and required little thinking. Students followed instructions and carried out investigation that was mostly of the fair testing type. This risks reducing the teaching of investigation to a "cook book" approach (Roth 1994), and learning to surface learning.

To sum up, it can be said that the teaching of science investigation was constrained, learning was limited, motivation to learn was largely extrinsic and reduced to achieving credits and grades.

5.2   Conclusion

The overall research focused on students' views about the three aspects of the study; *learning*, *motivation*, and *assessment*. These are reported in relation to the research questions below.

Research question:

1. *How does carrying out science investigation relate to students' learning and motivation to learn?*

   - Students said they "learnt" when the investigation led to "new" learning.
   - Students commented that they "learnt" when there was variety (a *carousel* of explorations rather than one activity for the entire lesson).

- Students learnt when what they were learning was relevant to everyday life.
- Students thought that hands-on practical work, irrespective of the type of investigation and whether within investigation or another practical activity, helped them learn science ideas.
- Students enjoyed doing hands-on practical work and were particularly interested in participating in practical work that had novelty and variety—especially the practical aspects of science investigation—regardless of the type of investigation.
- Students found report writing demotivating.
- Students expressed that repetition of practical work, including investigations they had experienced before, was demotivating.
- During practical work, including practical aspects of science investigation, students experienced opportunities to move around and work with their friends in a less formal learning environment. They found this motivational.
- Some students pointed out tasks that did not offer a cognitive challenge were demotivating.

2. *What affect does internal assessment have on year 11 students' learning and motivation to learn?*

- Students adopted a surface approach to learning investigation as fair testing and tended to rote learn answers to fulfil the requirement of assessment. There was an absence of evidence of learning exhibiting characteristics associated with a deeper approach to learning such as reflecting upon and evaluating the procedure.
- Students identified the extrinsic motivation of getting an Achieved or better grade for AS1.1 as the most common motivating factor for learning science investigation. This was associated with students setting narrow performance goals for learning science investigation.
- Assessment of the fair testing type of investigation for AS1.1 encouraged an emphasis on developing procedural knowledge (learning the steps to follow) rather than procedural understanding (knowing why they were following the steps).
- Low level learning outcomes were encouraged through assessment of science investigation as fair testing for AS1.1. Assessment requirements for the Achieved grade encouraged knowledge acquisition. Although the requirements for Merit and Excellence grades identify higher order thinking, this thinking is not necessarily used by students who commonly rote learn answers for assessment.

3. *How does the type of investigation relate to students' learning and motivation to learn?*

- Students said that hands-on practical work, irrespective of the type of investigation and whether within investigation or another practical activity, helped them learn science ideas.
- Students adopted a surface approach to learning investigation as fair testing and tended to rote learn answers to fulfil the requirement of assessment.

There was an absence of evidence of learning exhibiting characteristics associated with a deeper approach to learning such as reflecting upon and evaluating the procedure.

- Students enjoyed doing hands-on practical work and were particularly interested in participating in practical work that had novelty and variety—especially the practical aspects of science investigation—regardless of the type of investigation.
- Students found report writing demotivating.
- Students found repetition of practical work, including investigations they had experienced before, demotivating.
- During practical work, including practical aspects of science investigation, students experienced opportunities to move around and work with their friends in a less formal learning environment. They found this motivating.

The findings of the case study suggest that teaching, learning and motivation to learn science investigation were taken over by the requirements of internal assessment of science in year 11. That said, there are real insights offered by these students as noted above.

The research presented here has raised issues of validity and reliability of the internal assessment and recommends an evaluation of, and the need for, internal assessment for summative assessment in view of its unintended negative impact on learning.

References

Abrahams, I., & Millar, R. (2008). Does practical work really work? A study of the effectiveness of practical work as a teaching and learning method in school science. *International Journal of Science Education, 30*(14), 1945–1969. doi:10.1080/09500690701749305.

Alexander, P. A., Kulikowich, J. M., & Jetton, T. L. (1994). The role of subject-matter knowledge and interest in the processing of linear and nonlinear texts. *Review of Educational Research, 64*, 201–252.

Anderson, L. W., & Krathwohl, D. R. (Eds.). (2001). *A taxonomy for learning, teaching, and assessing: A revision of bloom's taxonomy of educational objectives.* New York: Longman.

Bloom, B., Englehart, M., Furst, E., Hill, W., & Krathwohl, D. (1956). *Taxonomy of educational objectives: The classification of educational goals. Handbook I: Cognitive domain.* New York: Longman.

Cleaves, A., & Toplis, R. (2007). Assessment of practical and enquiry skills: Lessons to be learnt from pupils' views. *School Science Review, 88*(325), 91–96.

Eccles, J. S., & Wigfield, A. (2002). Motivational belief, values, and goals. *Annual Review of Psychology, 53*, 109–132.

Entwistle, N. (2005). *Learning and studying: Contrasts and influences.* Retrieved September 8, 2005, from http://www.newhorizons.org/future/Creating_the_Future/crfut_entwi9stle.html.

Entwistle, N., & Ramsden, P. (1983). *Understanding student learning.* London: Croom Helm.

Fraser, B. J., McRobbie, C. J., & Giddings, G. J. (1995). Development and cross-national validation of a laboratory classroom environment instrument for senior high school science. *Science Education, 77*, 1–24. doi:10.1002/sce.3730770102.

Gott, R., & Duggan, S. (2002). Problems with the assessment of performance in practical science: Which way now? *Cambridge Journal of Education, 32*(2), 183–201.

Hall, C. (2007). *Qualitative research designs.* Wellington: School of Education Studies, Victoria University of Wellington (unpublished lecture notes).

Harlen, W. (2005). Trusting teachers' judgement: Research evidence of the reliability and validity of teachers' assessment used for summative purposes. *Research Papers in Education, 20*(3), 245–270.

Hodson, D. (1993). Rethinking old ways: Towards a more critical approach to practical work in school science. *Studies in Science Education, 22*, 85–142.

Hodson, D. (2014). Learning science, learning about science, doing science: Different goals demand different learning methods. *International Journal of Science Education.*

Hume, A., & Coll, R. (2008). Student experiences of carrying out a practical science investigation under direction. *International Journal of Science Education, 30*(9), 1201–1228.

Izard, C. E. (1977). *Human emotions.* New York: Plenum Press.

Keiler, L. S., & Woolnough, B. E. (2002). Practical work in school science: The dominance of assessment. *School Science Review, 83*(304), 83–88.

Kraska, M. (2008). Assessment. In N. J. Salkind (Ed.), *Encyclopedia of educational psychology.* Retrieved September 11, 2009, from http://www.sage-ereference.com/educationalpsychology/Article_n17.html.

Lederman, N. G., Antink, A., & Bartos, S. (2014). Nature of science, scientific inquiry, and socio-scientific issues arising from genetics: A pathway to developing a scientifically literate citizenry. *Science and Education, 23*(2), 285–302.

Millar, R. (2004). The role of practical work in the teaching and learning of science. In *Paper presented for the meeting of high school science laboratories: Role and vision.* Washington: National Academy of Sciences

Ministry of Education (1993). *Science in the New Zealand curriculum.* Wellington: Learning Media.

New Zealand Qualifications Authority. (2005). *Achievement Standard 1.1.* Retrieved June 24, 2009, from http://www.nzqa.govt.nz/ncea/assessment/search.do?query=Science&view=achievements&level=01.

Palmer, D. H. (2009). Students interest generated during an inquiry skills lesson. *Journal of Research in Science Teaching, 46*(2), 147–165.

Rennie, L. J., Goodrum, D., & Hackling, M. (2001). Science teaching and learning in Australian schools: Results of a national study. *Research in Science Education, 31*, 455–498.

Roberts, R. (2009). Can teaching about evidence encourage a creative approach in open-ended investigations? *School Science Review, 90*(332), 31–38.

Rotgans, J. I., & Schmidt, H. G. (2014). Situational interest and learning: Thirst for knowledge. *Learning and Instruction, 32*, 37–50.

Roth, W.-M. (1994). Experimenting in a constructivist high school physics laboratory. *Journal of Research in Science Teaching, 31*, 197–223.

Toplis, R. (2004). What do key stage 4 pupils think about science investigations? *Science Teacher Education, 41*, 8–9.

Watson, R., Goldsworthy, A., & Wood-Robinson, V. (1999). What is not fair with investigations? *School Science Review, 80*(292), 101–106.

Wellington, J. (1998). Practical work in science: Time for re-appraisal. In J. Wellington (Ed.), *Practical work in science. Which way now?* (pp. 3–15). London: Routledge.

Wellington, J. (2005). Practical work and the affective domain: What do we know, what should we ask, and what is worth exploring further? In S. Alsop (Ed.), *Beyond cartesian dualism* (pp. 99–109). Netherlands: Springer.

Final Thoughts

Internationally, most science curricula continue to mandate that students learn science content, do science, learn about the nature of science and develop the competencies to make informed decisions about the socio-scientific aspects relevant to their daily lives. Science education in the twenty-first century continues to emphasise the need to develop scientific literacy. Science education scholars acknowledge that school science investigation is different to the way scientists investigate due to the depth of conceptual and procedural understandings, as well as deep knowledge of the very nature of science investigation they have, compared to students. Furthermore, scholars assert that learning about the nature of scientific investigation (inquiry) is an achievable goal (Lederman et al. 2014).

Teachers believe that investigations in particular and practical work in general are motivational. Science investigative experiences can be memorable experiences as seen in the research presented here. Interest, enjoyment, and fun are emotions that can easily be experienced when students engage in investigation. Motivational scholars posit that interest is about being fascinated, curious, becoming engrossed and getting involved. Enjoyment is the satisfaction that comes from participating in an activity and the satisfaction of knowing that one is capable of coping with problems is sheer joy (Izard 1977; Ainsley and Hidi 2014).

Starting with diagnostic assessment to find out what the students already know. Ongoing formative assessment during learning to gauge what has been learnt, deciding on the next learning steps and scaffold learning, and then assess that learning is good practice. Focussing on what we *intend* the students to learn, using *evidence* to asertain whether this *intended learning* has taken place is the most important role of assessment in science (Millar 2013).

We can get students interested, provide opportunities for them to be curious, get satisfaction, and enjoyment from the confidence of knowing that they can find answers to their own questions through investigating. We can provide experiences that lead to new learning, deeper understanding, and appreciation of the relevance of what they are learning and its connectedness to everyday life. This can take place in most countries until about the age of 15 years, which appears to be the time when assessment begins to drive learning.

© The Author(s) 2015
A. Moeed, *Science Investigation*, SpringerBriefs in Education,
DOI 10.1007/978-981-287-384-2

If the focus of the first ten years of school science was *Beyond Play, and Assessment and on Learning through science investigation* it is likely that the students understand the nature of science investigation and will succeed in assessment without the need to be trained.

I believe it is a worthy aspiration for science education.

References

Ainley, M., & Hidi, S. (2014). Interest and enjoyment. In *International handbook of emotions in education*, pp. 205–227.

Izard, C. E. (1977). *Human emotions*. New York: Plenum Press.

Lederman, N. G., Antink, A., & Bartos, S. (2014). Nature of science, scientific inquiry, and socio-scientific issues arising from genetics: A pathway to developing a scientifically literate citizenry. *Science and Education, 23*(2), 285–302.

Millar, R. (2013). Improving science education: Why assessment matters. In *Valuing assessment in science education: Pedagogy, curriculum, policy* (pp. 55–68). Springer Netherlands.

Appendix 1
Achievement Standard 1.1

Subject Reference	Science 1.1			
Title	Carry out a practical science investigation with direction			
Level	1	Credits	4	Assessment	Internal
Subfield	Science			
Domain	Science—Core			
Registration date	27 October 2004	Date version published	27 October 2004

This achievement standard involves carrying out a practical investigation, with direction, by planning the investigation, collecting and processing the data, and interpreting and reporting the findings.

Achievement Criteria

Achievement	Achievement with Merit	Achievement with Excellence
• Carry out a practical science investigation	• Carry out a quality practical science investigation	• Carry out and evaluate a quality practical science investigation

Explanatory Notes

1. This achievement standard is derived from *Science in the New Zealand Curriculum*, Learning Media, Ministry of Education, 1993, 'Developing Scientific Skills and Attitudes', pp. 42–51; and *Pūtaiao i roto i te Marautanga o Aotearoa*, Learning Media, Ministry of Education, 1996, 'Ngā Pūkenga me Ngā Waiaro ki te Pūtaiao', pp. 70–85.

2. Procedures outlined in *Safety and Science: a Guidance Manual for New Zealand Schools*, Learning Media, Ministry of Education, 2000, should be followed. Investigations should comply with the Animal Welfare Act 1999, as outlined in *Caring for Animals: a Guide for Teachers, Early Childhood Educators, and Students*, Learning Media, Ministry of Education, 1999.

© The Author(s) 2015
A. Moeed, *Science Investigation*, SpringerBriefs in Education,
DOI 10.1007/978-981-287-384-2

3. An *investigation* is an activity covering the complete process: planning, collecting and processing data, interpreting, and reporting on the investigation. It will involve the student in the collection of primary data.

    The investigation will be directed. This means that general instructions for the investigation will be specified in writing and direction will be given in the form of the equipment and/or chemicals from which to choose. A template or suitable format for planning the investigation will be provided for the student to use.

4. Investigations should be based on situations in keeping with content drawn from up to and including science/pūtaiao curriculum Level 6. Possible contexts are given in the curriculum documents.

5. If a student enters for assessment against AS90186, Science 1.1, as well as any of: AS90156, Agriculture and Horticulture 1.1; AS90161, Biology 1.1; AS90169, Chemistry 1.1; or AS90180, Physics 1.1, the investigations must be in different subject areas. For example, if a student is being assessed against AS90161, Biology 1.1, and is also being assessed against AS90186, Science 1.1, then the emphasis of their investigation for AS90186, Science 1.1, cannot be based on biology.

6. A practical science investigation *will involve*:

    - a statement of the purpose—this may be an aim, testable question, prediction, or hypothesis based on a scientific idea
    - identification of a range for the independent variable or sample
    - measurement of the dependent variable or the collection of data
    - collecting, recording and processing data relevant to the purpose
    - a conclusion based on the interpretation of the processed data.

7. A quality practical science investigation *enables a valid conclusion to be reached. This would normally involve:*

    - a statement of the purpose—this may be an aim, testable question, prediction or hypothesis based on a scientific idea
    - a method that describes: a valid range for the independent variable or sample; a description of and/or control of other variables; the collection of data with consideration of factors such as sampling, bias, and/or sources of error
    - collecting, recording and processing of data to enable a trend or pattern (or absence) to be determined
    - a valid conclusion based on interpretation of the processed data that links to the purpose of the investigation.

8. Evaluate *means to justify the conclusion in terms of the method used. Justification will involve, where relevant, consideration of the:*

    - reliability of the data
    - validity of the method
    - science ideas.

Quality Assurance

1. Providers and Industry Training Organisations must be accredited by the Qualifications Authority before they can register credits from assessment against achievement standards.
2. Accredited providers and Industry Training Organisations assessing against achievement standards must engage with the moderation system that applies to those achievement standards.

Accreditation and Moderation Action Plan (AMAP) reference 0226.

Appendix 2
Achievement Standard Assessment
Task for AS1.1

National Certificate of Educational Achievement
TAUMATA MĀTAURANGA Ā-MOTU KUA TAEA

2008

Internal Assessment Resource

Subject Reference: **Science 1/1**

Internal assessment resource reference number:

Sci/1/1_CC2

Watch that car go

Supports internal assessment for:

Achievement Standard 90186 v3
Carry out a practical science investigation with direction

Credits: 4

Date version published:	April 2008
Ministry of Education	For use in internal assessment
quality assurance status	from 2008

© The Author(s) 2015
A. Moeed, *Science Investigation*, SpringerBriefs in Education,
DOI 10.1007/978-981-287-384-2

Teacher Guidelines:

The following guidelines are supplied to enable teachers to carry out valid and consistent assessment using this internal assessment resource. These teacher guidelines do not need to be submitted for moderation.

Context/setting:

This assessment resource is based on planning, carrying out, processing and interpreting, and reporting of a practical investigation that is a **fair test** investigation. The teacher directs what type of investigation the students are to do and changes the planning sheets and student instructions accordingly.

Conditions:

This assessment activity is to be carried out in four parts that lead to the production of an investigation report.

The specific conditions should be stated on the student instruction sheet. e.g. equipment and materials available.

The students need sufficient time for:

- Trialling and planning
- Carrying out
- Processing and interpreting data
- Writing a report

The time allowed will depend on the particular investigation chosen. State this time on the student instruction sheet.

Any special safety requirements **must** be stated on the student instruction sheet.

Teachers need to be aware of the credit value of this standard when determining the time needed to carry out the investigation.

Resource requirements:

Students will need to be provided with the materials and equipment required for trialling and carrying out the investigation.

The Investigation

Part 1: Developing a Plan

- The student is provided with a *planning Sheet* (included) and will work independently to complete this. The planning sheet may need to be modified, related to the task chosen, to allow sufficient space for students to write.
- The student should be given the opportunity to conduct trials to develop their method, eg. to establish a suitable range of values for the independent variable for a fair test or the sample selection for pattern seeking. A record of this trialling needs to be mentioned on the template or in the final report.
- The student uses the planning sheet and trial results to write a detailed, step-by-step method. The Planning sheet (or other check sheets) may be used to self-evaluate that the method is workable.

Part 2: Collecting and Recording Data

The student follows their written method to collect their own data. The method may be modified but these modifications must be included in their final report and indicated to the assessor.

Part 3: Processing and Interpreting Results

The student must process the data collected into a form that shows a pattern or a trend or absence. This may be achieved by averaging, using a table or using a graph.

Part 4: Presenting a Report

The student, working independently, presents the report of the investigation following the directions/format given in the student instructions.

Teacher Resource Sheet

Prior teaching will need to occur on the scientific method and how to design a practical that involves **'fair' testing**.

2008

Internal Assessment Resource

Subject References: **Science 1.1**
Internal assessment resource reference number: Sci/1/1_CC2

Watch that car go

Supports internal assessment for:
Achievement standard 90186 v3
Carry out a practical science investigation with direction
Credits: 4

Student Instructions Sheet	
School/Institution	
Student Name	
Teacher or Class reference	
Date of completion	

Background Information:

Watching children play with their toy cars student noticed that the cars seemed to go faster and further the sleeper the slope they were released on.

> In this investigation you are to develop and carry out an investigation. You will plan, collect, process and interpret information, and present a report to find out how the slope of a ramp affects the distance the car goes along the flat(table or floor) This investigation is a **fair test** investigation.

Conditions:

This assessment activity is to be carried out in four parts that leads to the production of an investigation report. This investigation must be carried out individually.

Times: Total time will take a minimum of 1 week e.g.

• trialling and planning	1 period
• carrying out and processing data	1 period
• interpreting the data and	1 period
• writing the final report	1 period

Equipment:

You have been given books, a ruler, ramp, and a car.

Part 1: The plan

1. State the purpose of your investigation
2. Identify the key variables involved:

   • the independent variable (the variable that is to be changed)
   • the dependent variable (the variable that will be measured)
   • controlled variables (significant or relevant variables that will need to be kept the same to make your results more reliable)

3. Describe a suitable range of values to be used for the independent variable and how these values will be changed. Trialling will help you establish this range.
4. Describe how the dependent variable will be measured.

Controlled variables: (things you will need to keep the same)

1. Identify any other variables that might influence your investigation and describe how they will be controlled or kept the same to make your results more accurate.
2. Describe how you will ensure that your results are reliable and that you have enough data.

Now write a detailed **step-by-step method** that you will use.

You may change your method as you carry it out as long as you describe any changes made to the method in your report.

Part 2: Collect and Record Data

• Follow your method to collect data and record the results in a table or another appropriate way.
• Remember to record any changes to your method and reasons for the changes as you go.
• Record any difficulties with equipment, gathering your data or your method.

Part 3: Process and Interpret Results

• Process your results so that you can show the trend (or lack of) or pattern in your data. This will usually involve some calculations (e.g. averages) and/or a graph.
• Record the relevant trend or pattern; this is your interpretation.
• Relate the trend or pattern to your purpose; this is your conclusion.

Part 4: Present a Report

Present a report on your investigation. This will include your:

- Trialling and planning sheet
- Detailed step-by-step method, including any changes made during your investigation
- Recorded data
- Processed data
- Interpretation of results
- Conclusion that links your interpretation to the purpose of the investigation evaluation of the conclusion in terms of the method used. In this you may comment on the

  – Reliability of data (repeats/outliers etc.)
  – Validity of the final method chosen.
  – Way your results reflect the science ideas related to the investigation.

Planning Sheet   Student name:

1. Purpose of investigation (this may be an aim, testable question, prediction or hypothesis)
2. **FAIR TEST**   Which variable will be changed? (This is the independent variable)   How will the independent variable be changed?   Give a suitable range of values for this variable
3. **FAIR TEST**   Which variable will have to be measured or observed in order to get some data or information from the investigation? (This is the dependent variable)   How will the dependent variable be measured or observed?

4. Other variables that need to be controlled to make your results more accurate.	
Other Variables	Describe how this variable will be controlled or kept the same?

5. How will you ensure that your results are reliable?
6. Notes from your trials.

Report Sheet

Recorded data:

Processed data:

Interpretation of Data:

Conclusion:

Evaluation of the Method and Data and / or Science ideas

Assessment Schedule: Sci/1/1_CC2 - Watch that car go

Evidence	Achievement	Merit	Excellence
Report contains	Statement of purpose Eg: To see what effect the height of a ramp above the floor has on the distance travelled by a car.	(as for achievement)	(as for achievement)

Report (planning sheet and or method)	Identify range for independent variable.
	
	• *The height of the ramp above the floor. Only 3 heights chosen.*
	
	and
	• Measurement of dependent variable.
	Distance travelled from the ramp.
		a valid method (easily followed by another student) is required that includes:
		
		• A **valid** range of the independent variable.
		  Eg: At least 4 independent variables (height), preferably 5, is the minimum recommended range for Merit grade.
		  Eg: heights of 5cm, 10cm, 15cm, 20cm
		• Description of **and** control of other key variables
		  Eg: same ramp, same position released on the ramp, same area of bench / floor, same car.
		• Consideration of other factors
		  Eg: how the car released.
		
		and
		• Measurement of dependent variable.
		  Distance travelled from the ramp in cm.
			As for merit

| Report (Recorded data, processed results) | Collect, record and process data relevant to purpose.

one investigation on each height. Distance recorded in follow able format. Graph showing trend.

<Possible results>
<table><tr><td>Height (cm)</td><td>Distance travelled (cm)</td></tr><tr><td>5</td><td>3.4</td></tr><tr><td>10</td><td>12.3</td></tr><tr><td>15</td><td>33.5</td></tr><tr><td>20</td><td>46.0</td></tr></table> | Collect, record and process data to enable a **valid** pattern or trend (or absence)

More than two repeats (to establish validity) on chosen heights. Results correctly averaged

Distance accurately measured.

Averages recorded in a way that allows the **valid** trend to be shown and/or correctly labelled graph-showing **valid** trend that supports collected data. | (as for merit) |
| Report (Interpretation and conclusion) | A conclusion based on the processed data collected.

The greater the height of the slope the greater the distance the car travelled. | A **valid** conclusion that links to the purpose.

The greater the height of the ramp and therefore the steeper the slope the greater the distance travelled by the car. | (as for merit) |

| Report (evaluation) | | | **Evaluation to justify the method used to reach the conclusion.**
One of
Method
The final method chosen gave results that were repeatable. This also allowed us to see aberrant data and this data could be removed from our calculations.
Reliable data.
We took enough reading to allow a valid trend. We repeated each trail until we got 3 recordings that were similar.

Science ideas
Example
Our results supported the science idea that
The higher the ramp the more gravitational potential energy the car had and the more that was converted to kinetic energy, therefore the further the car went before it came to a halt. |

To determine the overall level of performance all judgements within a column must be met.
For each judgement, evidence can be obtained from anywhere in the report.